Sanyuan Jiang

Hydrological water quality modelling of nested meso scale catchments

Sanyuan Jiang

Hydrological water quality modelling of nested meso scale catchments

Südwestdeutscher Verlag für Hochschulschriften

Impressum / Imprint
Bibliografische Information der Deutschen Nationalbibliothek: Die Deutsche Nationalbibliothek verzeichnet diese Publikation in der Deutschen Nationalbibliografie; detaillierte bibliografische Daten sind im Internet über http://dnb.d-nb.de abrufbar.
Alle in diesem Buch genannten Marken und Produktnamen unterliegen warenzeichen-, marken- oder patentrechtlichem Schutz bzw. sind Warenzeichen oder eingetragene Warenzeichen der jeweiligen Inhaber. Die Wiedergabe von Marken, Produktnamen, Gebrauchsnamen, Handelsnamen, Warenbezeichnungen u.s.w. in diesem Werk berechtigt auch ohne besondere Kennzeichnung nicht zu der Annahme, dass solche Namen im Sinne der Warenzeichen- und Markenschutzgesetzgebung als frei zu betrachten wären und daher von jedermann benutzt werden dürften.

Bibliographic information published by the Deutsche Nationalbibliothek: The Deutsche Nationalbibliothek lists this publication in the Deutsche Nationalbibliografie; detailed bibliographic data are available in the Internet at http://dnb.d-nb.de.
Any brand names and product names mentioned in this book are subject to trademark, brand or patent protection and are trademarks or registered trademarks of their respective holders. The use of brand names, product names, common names, trade names, product descriptions etc. even without a particular marking in this work is in no way to be construed to mean that such names may be regarded as unrestricted in respect of trademark and brand protection legislation and could thus be used by anyone.

Coverbild / Cover image: www.ingimage.com

Verlag / Publisher:
Südwestdeutscher Verlag für Hochschulschriften
ist ein Imprint der / is a trademark of
OmniScriptum GmbH & Co. KG
Heinrich-Böcking-Str. 6-8, 66121 Saarbrücken, Deutschland / Germany
Email: info@svh-verlag.de

Herstellung: siehe letzte Seite /
Printed at: see last page
ISBN: 978-3-8381-5003-1

Zugl. / Approved by: Braunschweig,Technischen Universität Carolo-Wilhelmina zu Braunschweig,Dissertation,2014

Copyright © 2015 OmniScriptum GmbH & Co. KG
Alle Rechte vorbehalten. / All rights reserved. Saarbrücken 2015

Acknowledgment

The objectives of this research project would not have been met without the support of several people. Thus, here I would like to express my gratitude for many interesting comments and suggestions as well as for the encouragement I have received during my thesis in connection with the data analysis and the model application.

Most importantly, I would like to thank my supervisor PD Dr. Michael Rode from Department Aquatic Ecosystem Analysis and Management, Helmholtz Centre for Environmental Research-UFZ for his strong encouragement, support and especially for his endless patience in discussing the results and improving my writing. His guidance, remarks and critical comments on my work have played a crucial role in both the contents and presentation of this thesis. Moreover, the regular intensive scientific discussion significantly improved my physical understanding of the water and nutrient transport processes at catchment scale and helped foster the development of a scientific attitude in my work. Finally, I am deeply thankful to Dr. Rode for his continual help, understanding and for his belief in me.

I would also like to give thanks to my supervisor Professor Günter Meon from Leichtweiß-Institute for Hydraulic Engineering and Water Resources, Technische Universität Braunschweig for his expertise and interest in my work and support on my thesis. His useful comments and feedbacks on the results of this study were beneficial to the project. Additionally, I want to express my thanks to Professor Dietrich Borchardt from Department Aquatic Ecosystem Analysis and Management, Helmholtz Centre for Environmental Research-UFZ for his interest and support on my work.

For excellent collaboration and scientific discussions and, even more, for their friendship, I would like to thank my colleagues Dr. Seifeddine Jomaa and Dr. Olaf Büttner. I shall not forget my colleagues at the group hydrological and water quality modelling for their technical support and guidance. Moreover, I would like to extend my gratitude to the secretary, Martina Klapputh, for the facilities and helps that she gave me during my study.

I would also like to thank my colleagues Philipp Theuring and Melanie Hartwig, Dr. Muhammad Rehan Anis, Gundula Paul, Andrew Kaus, Julia Vanessa Kunz for their friendship and the constructive atmosphere created in the office. I shall not forget my colleague and friend Dr. Sumit Sinha for the interesting discussions and help on improving my English writing.

Finally, I would like to express my thanks to my family and my wife's family, especially to my wife Xia Chen for their consideration and support on my life.

Tables of contents

Acknowledgment ii

Zusammenfassung v

Abstract vii

Chapter 1: Introduction 1

1.1   Problem statement 1

1.2   State of art 2

    1.2.1 Hydrological and nutrient transport processes 2

    1.2.2 Hydrological water quality modeling 5

    1.2.3 Applicability of hydrological water quality models 8

    1.2.4 Model calibration and uncertainty analysis 9

    1.2.4.1 Uncertainty sources 9

    1.2.4.2 Model calibration and uncertainty analysis methods 13

    1.2.5 Step-wise calibration vs. multi-objective calibration 17

    1.2.6 Effects of spatial and temporal resolution of calibration data on model identification 18

1.3   Knowledge gaps 20

1.4   Objectives 22

1.5   Structure of Dissertation 22

Chapter 2: Materials and Methodologies 25

2.1   Selke catchment 25

2.2   Weida catchment 29

2.3   Hydrological water quality model (HYPE) 31

    2.3.1 Hydrological processes 33

    2.3.2 Nitrogen processes 36

    2.3.3 Input data 38

    2.3.4 Model parameter set and parameterization 39

2.4   Model calibration and uncertainty analysis approaches 41

    2.4.1 PEST 41

    2.4.1.1 Parameter estimation 41

    2.4.1.2 Sensitivity analysis 41

    2.4.1.3 Predictive analysis 42

2.4.2 DREAM$_{(ZS)}$ 42
2.5   Model setup and evaluation 44
  2.5.1 HYPE model set up 44
    2.5.1.1 Model setup at Selke catchment 44
    2.5.1.2 Model setup at Weida catchment 44
  2.5.2 Model performance evaluation 45
    2.5.2.1 Inorganic nitrogen loads calculation 45
    2.5.2.2 Model performance evaluation criteria 45
2.6   Multi-site and multi-objective calibration 47
2.7   Step-wise calibration and multi-objective calibration 47
2.8   Effects of spatial and temporal resolution of calibration data on model identification .. 48
  2.8.1 Single-site vs. Multi-site calibration 48
  2.8.2 Bi-weekly vs. Daily IN measurements 49

Chapter 3: Multi-site and multi-objective calibration of integrated catchment model HYPE 51
3.1   Model calibration, hydrological and IN simulation results at Selke catchment 51
  3.1.1 Model parameter calibration results 51
  3.1.2 Hydrological simulation 53
  3.1.3 IN simulation 57
    3.1.3.1 IN concentration and daily IN load 57
    3.1.3.2 Monthly and yearly IN loads simulations 60
3.2   Discussion 63

Chapter 4: Comparison of step-wise and multi-objective calibration 67
4.1   Modeling results from SWC and MOC at Weida catchment 67
  4.1.1 Model parameter calibration results 67
    4.1.1.1 Parameter sensitivity 67
    4.1.1.2 Parameter optimized values and posterior uncertainty 70
  4.1.2 Hydrological simulation 70
  4.1.3 IN simulation 72
    4.1.3.1 IN concentration and daily IN load 72
    4.1.3.2 Monthly and annual IN loads simulation 75
4.2   Modeling results from SWC and MOC at Selke catchment 77

4.2.1 Model parameter calibration results 77
4.2.2 Model simulation results 78
Chapter 5: Effects of spatial and temporal resolution of calibration data on model identification 81
5.1   Effects of spatial resolution of calibration data 81
  5.1.1 Discharge simulation 81
    5.1.1.1 Calibration and predictive analysis using PEST 81
    5.1.1.2 Calibration and predictive analysis using DREAM$_{(ZS)}$ 92
  5.1.2 IN concentration simulation 98
    5.1.2.1 Calibration and predictive analysis using PEST 98
    5.1.2.2 Calibration and predictive analysis using DREAM$_{(ZS)}$ 104
5.2   Temporal resolution effect of calibration data 109
  5.2.1 Discharge simulation 110
  5.2.2 Calibration and predictive analysis using PEST 111
  5.2.3 Calibration and predictive analysis using DREAM$_{(ZS)}$ 120
5.3   Comparison of PEST and DREAM$_{(ZS)}$ 125
Chapter 6: Conclusions 127
References 129
Appendix 1: Hydrological nitrogen processes in the HYPE model 140
  Appendix 1A: Model equations 140
  Appendix 1B: Notation 144
Appendix 2: How PEST does optimization, sensitivity analysis and predictive analysis 148
  Appendix 2A: parameter optimization 148
  Appendix 2B: Sensitivity analysis 148
  Appendix 2C: Predictive analysis 149
Appendix 3: Bayesian inference of the posterior probability density function of hydrologic model parameters using DREAM$_{(ZS)}$ 151

Zusammenfassung

Hydrologische Gewässergütemodellierung ist ein immer wichtiger werdendes Werkzeug zur Untersuchung von Abfluss- und Nährstofftransportprozessen und kann wichtige Erkenntnisse für ein angewandtes Einzugsgebietsmanagement liefern. Das HYPE Model (HYdrological Predictions for the Environment) ist ein semi-distributives, hydrologisches Wassergütemodel, das Abfluss und Nährtstoffkonzentrationen (N und P) auf Tagesbasis auf Einzugsgebietsebene berechnen kann. Das Model wurde in Schweden entwickelt und erstmals erfolgreich angewandt, wurde aber bisher noch nicht in anderen Einzugsgebieten mit unterschiedlichen physiogeografischen, klimatischen, hydrologischen Bedingungen und Gewässerbelastungen eingesetzt. In dieser Studie wurde das HYPE Model zur Simulation von Abfluss und Konzentration von inorganischem Stickstoff (IN) für zwei Einzugsgebiete der unteren Mittelgebirgslagen, der Selke (463 km^2) und der Weida (99 km^2), getestet. Evapotranspiration und IN-Senken (Pflanzenaufnahme und Denitrifikation) sind die sensitivsten Prozesse für die Simulation von Abfluss und Nitrat. Die Ergebnisse zeigten, dass sich die IN-Konzentration und die tägliche IN-Fracht proportional zum Abfluss verhalten. Dies deutet darauf hin, dass der Abbau von IN in anthropogen genutzen Einzugsgebieten vom Abfluss abhängig ist. Das HYPE Model konnte die Dynamiken von Abfluss und IN-Konzentrationen gut abbilden (Nash-Sutcliffe Koeffizient >0.83). Der Vergleich zwischen schrittweiser Kalibrierung (d.h. die Kalibrierung der hydrologischen Parameter erfolgt vor der Kalibrierung der Gewässergüteparameter) und multiobjektiver Kalibrierung (simultane Kalibrierung der hydrologischen und Gewässergüteparameter) zeigt, dass die Modellgenauigkeit bei multiobjektiver Kalibrierung höher und robuster ist. Dies weist darauf hin, dass die multiobjektive Kalibrierung eine bessere Identifikation von Parametern in hydrologischen Gewässergütemodellen erlaubt, da hydrologische Prozesse und Gewässergüteprozesse korrelieren.

Zur Parameterkalibrierung und Unsicherheitsanalyse wurde das HYPE Modell mit den Tools PEST (Model-Independent Parameter Estimation & Uncertainty Analysis) und DREAM$_{(ZS)}$ (DiffeRential Evolution Adaptive Metropolis Algorithm) kombiniert. PEST ist eine modellunabhängige Software für die Parameterschätzung und Unsicherheitsanalyse, die den Gauss-Marquardt Algorithmus verwendet. DREAM$_{(ZS)}$ ist ein Markov Chain Monte Carlo (MCMC) Such- und Optimierungsalgorithmus für die Bayesische Inferenz der posteriori-Wahrscheinlichkeit hydrologischer Modellparameter. „Multi-site calibration" (Parameteroptimierung anhand von Daten mehrerer Messstellen im Einzugsgebiet) wurde mit „Single-site calibration" (Parameteroptimierung anhand von Daten nur einer Messstelle) verglichen. Durch eine "Multi-site calibration" konnten die Modellgenauigkeit für gebietsinterne Messstellen erhöht sowie die Vorhersageunsicherheiten vermindert werden. Dies belegt die Bedeutung der Beobachtungen von gebietsinternen Messstellen für die räumlich verteilte Vorhersage. Dies kann damit erklärt werden, dass räumliche Heterogenitäten der Einzugsgebietscharakteristika bei einer „Multi-site calibration" besser berücksichtigt werden. Die Kalibrierung von Stickstofftransport- und -umsatzprozessen mittels Nitrattageswerten führt im Vergleich zur Verwendung von 2-wöchentlichen Werten zu einer höheren Modellgenauigkeit. Zudem werden Parameterunsicherheiten verringert. Dies ist darauf zurückzuführen, dass die Variabilität hydrologischer Bedingungen durch zeitlich hoch aufgelöste Messdaten besser erfasst werden. Sowohl PEST als auch DREAM$_{(ZS)}$ erwiesen sich als geeignet zur Kalibrierung der Modellparameter. Allerdings weist DREAM$_{(ZS)}$ Vorteile gegenüber PEST auf, da es über einen globalen Suchalgorithmus verfügt und Vorhersageunsicherheiten anhand Bayesischer Inferenz objektive berechnet werden.

Abstract

Hydrological water quality modeling is increasingly used for investigating runoff and nutrient transport processes as well as providing guidelines for watershed management. The HYPE (HYdrological Predictions for the Environment) model is a semi-distributed hydrological water quality model that simulates streamflow and nutrient (N and P) concentrations on a daily time step at catchment scale. It was developed and applied successfully in Sweden but not intensively tested in other regions that show different physiographic, climatic, hydrological and water quality conditions. In this study, the HYPE model was tested for simulation of discharge and streamwater inorganic nitrogen (IN) concentration in two different mesoscale catchments of the German lower mountain range, the Selke (463 km^2) and Weida catchments(99 km^2). Evapotranspiration and IN sinks (plant uptake and denitrification) are found to be most sensitive processes for runoff and nitrogen simulations, respectively. Results showed that IN concentration and daily IN load had a proportional relationship with discharge, indicating that IN leaching is mainly controlled by runoff in managed catchments.The HYPE model was proved to be capable of capturing dynamics and balances of water and IN load with a Nash-Sutcliffe coefficient above 0.83. Through comparing step-wise calibration (calibrating hydrological parameters first and then water quality parameters) and multi-objective calibration (calibrating hydrological and water quality parameters simultaneously), it was found that the model performance obtained using the parameter set optimized from multi-objective calibration is better and more robust. This indicates that multi-objective calibration is more appropriate for parameter identification of integrated hydrological water quality models because both processes are correlated.

PEST (Model-Independent Parameter Estimation & Uncertainty Analysis) and DREAM$_{(ZS)}$ (DiffeRential Evolution Adaptive Metropolis algorithm) were combined with the HYPE model to implement parameter calibration and uncertainty analysis. PEST is a model-independent software for parameter estimation and uncertainty analysis using the Gauss-Marquardt algorithm. DREAM$_{(ZS)}$ is a Markov chain Monte Carlo (MCMC) algorithm for Bayesian inference of the posterior probability density function of hydrologic model parameters. Multi-site calibration (parameter optimization using measurements from both catchment outlet and internal sites) was compared with single-site calibration (parameter optimization using measurements only from catchment outlet) on model identification. Results showed that multi-site calibration improved model performances at internal sites and decreased parameter posterior uncertainty ranges and prediction uncertainty, indicating the importance of observations from internal sites for spatially distributed prediction. This can be explained by the fact that spatial heterogeneity of catchment characteristics are accounted for under multi-site calibration. Compared with the parameter calibration against biweekly nitrate-N concentration measurements, nitrogen-process parameters calibrated using daily averages of nitrate-N concentration observations produced better and more robust model performance on simulations of IN concentration and IN load, narrower posterior parameter uncertainty ranges and IN concentration prediction uncertainty. This is attributed to the fact that different hydrological conditions are covered under a temporal high resolution monitoring program. Both PEST and DREAM$_{(ZS)}$ are found to be efficient for hydrological water quality parameter calibration. However, DREAM$_{(ZS)}$ is more sound and appropriate than PEST because of its capability to evolve parameter posterior probability density functions and estimate prediction uncertainty objectively based on Bayesian inference.

Chapter 1: Introduction

1.1 Problem statement

Freshwater is an indispensable natural resource for survival of human-being and other species. The quantity and quality of freshwater influence human life, society stability and economy development. The increasing population and intensity with which land is used for crop production is reflected in changed land's surface and higher nutrient concentrations in many rivers and lakes (Heathwaite, 1995; Smith et al., 1999). Eutrophication is an accelerated growth of algae on higher forms of plant life caused by the enrichment of water by nutrients, especially by compounds of nitrogen and/or phosphorus.Eutrophication causes undesirable disturbance to the balance of organisms present in the water and to the quality of the water concerned. Streamwater eutrophication is widely reported to be a serious aquatic problem in many countries due to physical alterations affecting the hydrology and/or geomorphology of a water body and nutrient inputs from anthropogenic sources, which deteriorates the water quality and ecological services. Some examples of the problems caused by eutrophication are algal blooms, "red tides", "green tides", fish kills, inedible shellfish, blue algae and public health threats. The sources of streamwater eutrophication include nitrogen and phosphorus coming from diffuse and point sources, of which diffuse sources, mainly from agricultural activities, are more important (Lam et al., 2012, Rode et al., 2008). It was reported that nitrogen accounted for approximately 80% of all inputs, while phosphorus accounted for approximately 70% from 1998 to 2000 in Germany (BMU, 2005). Diffuse sources include dry and wet atmospheric deposition, manures, plant residues and nutrient from fertilizer inputs. Point sources include wastewater effluent (municipal and industrial) and overflows of combined storm and sanitary sewers etc. (Smith et al., 1999).

The currently legislated European Water Framework Directive (WFD) aims to achieve a 'good status' for all surface waters in European Union states by 2015 (EC, 2000; Rode et al., 2008).'Good status' means 'good ecological status' and 'good chemical status'. Good ecological status is defined in terms of the quality of the biological community, the hydrological characteristics and the chemical characteristics. Good chemical status is defined in terms of compliance with all the quality standards established for chemical substances at European level. Many studies revealed a close relationship between nutrient (N and P) in the rivers/lakes and benthic algae growth, algal biomass, planktonic chlorophyll, autotrophic and heterotrophic activities in rivers/lakes and biotic integrity (e.g. Dodds, 2006; Van Nieuwenhuyse and Jones, 1996). The deleterious effect of increased nutrient concentrations on fish communities in low order streams was noted when nutrient concentrations exceeded background conditions (total inorganic nitrogen and phosphorus > 0.61 mg l^{-1} and 0.06 mg l^{-1}, respectively) (Miltner and Rankin, 1998). These findings confirm the importance of non-point sources of pollution in catchment planning as well as the combined effect of habitat and riparian quality on nutrient assimilation. Although several measures have been taken to reduce nutrient inputs (e.g. reduction of fertilizer application), nutrient concentrations in water bodies are still unduly high.

In Germany, the nitrogen pressure on the environment is caused by mineral fertilizer, livestock manure, atmospheric deposition and biological fixation with share of 49.3%, 35.5%, 13.7% and 1.5%, respectively (European commission report, 2002). River basin management is an interdisciplinary task and includes components from both the natural sciences (hydrology, erosion and sediment transport, landscape assessment, hydrogeology etc.) and social sciences (socio-economics, ecological economics, behavioral theory etc.) (Rode *et al.*, 2002). To investigate the causes and regimes of eutrophication, several projects were implemented in Germany at selected river basins (e.g. Weiße Elster, Havel, Werra, Ems). From 2008, the infrastructure activity TERENO (TERrestrial ENvironmental Observatories) was started to establish an observation platform linking terrestrial observatories in different regions in Germany (http://teodoor.icg.kfa-juelich.de/overview-de, last accessed 30 August 2012). TERENO is embarking on new paths with an interdisciplinary and long-term research program involving six Helmholtz Association Centers. TERENO spans an earth observation network across Germany that extends from the North German lowlands to the Bavarian Alps. This unique large-scale project aims to investigate the long-term ecological, social and economic impact of global change at regional level. As one of four "Global Change Exploratories" planned by the Helmholtz Association, the UFZ (Helmholtz Centre for Environment Research) observatory focuses on the monitoring, analyzing and prediction of changing state variables and fluxes in terrestrial systems (http://www.ufz.de/index.php?en=16350, last accessed 30 August. 2012). The Bode river catchment, which spans the Harz mountain range to lowland areas in central Germany, was chosen as the observatory site in middle part of Germany. Current surveys conducted in the course of the implementation of the WFD suggest that 76% of the Bode river system will unlikely or uncertainly attain a good ecological status (49% unlikely while 27% uncertainly) because of high nutrient loads and heavily modified river morphology (Landesbericht, 2005).

1.2 State of art

1.2.1 Hydrological and nutrient transport processes

Nutrient leaching from landscapes to stream involves complex hydrological transport and nutrient turnover processes (Basu *et al.*, 2011; Onderka *et al.*, 2012). The schematic structure of nutrient (nitrogen and phosphorus) transport and transformation from hillslope to stream is shown in Figure 1.1. Hydrological processes are found to be dominant in controlling nutrient export, especially the dissolved nitrate component (Li et al., 2010; van Griensven et al., 2006). Nitrogen and phosphorus have different transport and transformation processes. Phosphorus is mainly transported as particulate and soluble P through overland flow due to its strong binding to soil particles. The most important processes for the transport of phosphorus from diffuse sources are soil erosion and surface runoff (Jiang and Rode, 2012; Rode and Lindenschmidt, 2001; Kistner et al., 2013). While, nitrogen is primarily leached from soil column and transported in dissolved forms by subsurface flow. Therefore, nitrogen concentrations of streamwater are mainly controlled by shallow subsurface flow (Lam et al., 2012; Liu et al., 2008). This can be explained as follows: (i) nitrogen is mainly stored in the topsoil and its amount decreases along the soil profile, and (ii) the shallow subsurface flow has short transport time, which causes low retention (e.g. denitrification) (Hesser et al., 2010). Deep groundwater flow has a long residence time, which results in high nutrient retention (Hesser et al., 2010; Wriedt and Rode, 2006). Due to hydrological and ecological interactions, relatively constant nutrient concentrations were found in some human-impacted agricultural regions (Li et al., 2010). A proportional relationship between nutrient emission load and discharge was found in many managed catchments, which indicates that nutrient leaching is transport-limited rather than supply-limited (Basu et al., 2010; Shrestha et al., 2012; Zhang, 2011). This enables the nutrient emission load to be calculated using a simple regression model with monitored discharge. However, a process-based hydrological water quality model of proper complexity is required to investigate internal nutrient processes (such as denitrification and plant uptake), effects of climate and land use changes on hydrological regimes, nutrient transport and turnover.

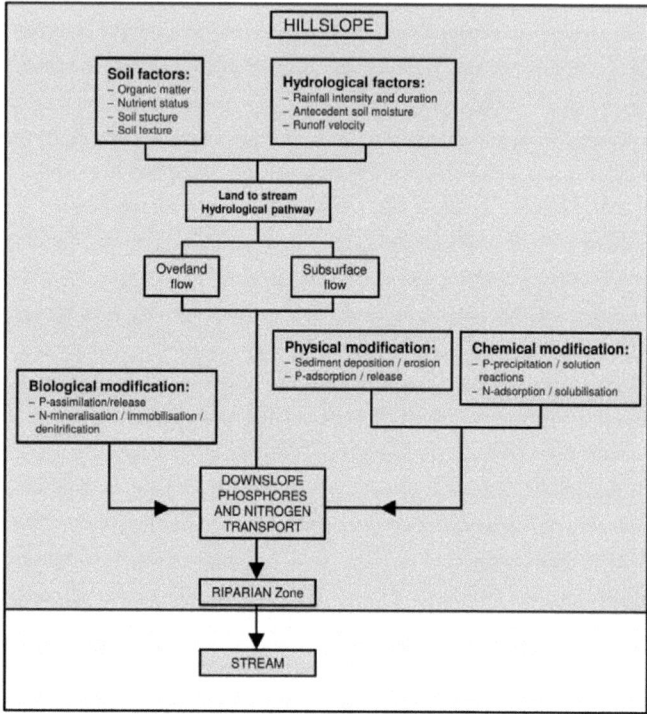

Figure 1.1 Nutrient (N and P) transport and transformation from hillslope to stream (after Heathwaite, 1995).

Hydrological processes are highly dependent on climate patterns (e.g. precipitation and temperature), topography, soil property, geological conditions and concerned catchment scale (Liu et al., 2008). As different flow pathways are the main driving forces of nutrient leaching, climate patterns and catchment physiographical characteristics also influence nutrient emission from landscapes to the stream. More nitrogen was found to be transported into stream in catchment areas covered with highly permeable soils (like sandy soil) compared with less permeable soil (e.g., clay soil) due to greater transport capacity related to higher hydrological conductivity and lower retention amounts attributed to short residence time (Askegaard et al., 2005; Kyllmar et al., 2006; Rode et al., 2009, Shrestha et al., 2013). Land use and agricultural practices have great effects on stream nutrient loads (e.g. Hansen et al., 2001; Rode et al.,

2009and 2010). It was reported that the nutrient concentrations in the streams were highly correlated with the percentage of agricultural land, as more agricultural land contribute to more fertilizer loss (Lagzdins *et al.*, 2012). As climate variability (precipitation and temperature) results in changes of hydrological regimes and nutrient turnover processes (e.g. plant uptake and denitrification due to temperature effects) (Arnell, 1999; Andersen *et al.*, 2006), it influences nitrate leaching from landscapes to stream in turn (Bouraoui *et al.*, 2004; De Klein and Koelmans, 2011; Li *et al.*, 2010).

1.2.2 Hydrological water quality modeling

Reliable methods for the estimation of various sources' contributions to the streamwater eutrophication as well as methods that predict the potential outcomes from climate change and watershed management scenarios are needed for environment policy making and successful implementation. Hydrological water quality modeling is increasingly used to investigate nutrient leaching and global change effects on riverwater quality. Simulation of climate change and watershed management scenarios requires a process-based water quality model, which should be linked to a hydrological model.Over the past decades, several integrated hydrological water quality models have been developed and tested on different catchment scales all over the world. Some popularly used hydrological water quality models are HSPF (Hydrological Simulation Program-Fortran) (Bicknell *et al.*, 2001), SWAT (Soil and Water Assessment Tool) (Arnold *et al.*, 1998), SWIM (Soil and Water Integrated Model) (Krysanova *et al.*, 2005), LASCAM (Large-scale catchment model) (Viney *et al.*, 2000), INCA (Whitehead *et al.*, 1998 a) and HYPE (Hydrological Predictions for the Environment) (Lindström *et al.*, 2010).

HSPF is a process-based catchment hydrological water quality model, which calculates time series of runoff, sediment load, nutrient and pesticide concentrations based on time series of rainfall, temperature and solar radiation as well as land use pattern and land management practices. SWAT is a process-based semi-distributed hydrological water quality model, which predicts the hydrologic cycle, movement of pesticides, sediments and nutrient based on climate data of daily precipitation, maximum/minimum air temperature, solar radiation, wind speed and relative humidity and spatial data of land use, soil cover and land management practices. SWIM simulates runoff generation, nutrient and carbon cycling, river discharge, plant growth, crop yield and erosion on mesoscale catchment based on climate, land use and soil. LASCAM was developed to simulate water quantity and water quality (represented by salt, sediments and nutrients) at large scale catchments. The HYPE model was developed by the Swedish Meteorological and Hydrological Institute (SMHI) between 2005-2007. It is a process-based, semi-distributed hydrological water quality model, which was developed based on the hydrological model HBV (e.g. Bergström, 1976; Lindström *et al.*, 1997) and the water quality model HBV-NP (e.g. Andersson *et al.*, 2005; Arheimer *et al.*,

2005; Lindström et al., 2005). HYPE predicts streamflow, streamwater nutrient (N and P) concentrations at daily time step based on climate data of precipitation and temperature, soil, land use and land management practices. The INCA model is a process-based and semi-distributed hydrological water quality model, which simulates nitrogen leaching at watershed scale on daily time step. Nitrogen transformations in both soil and in-stream systems (e.g. plant uptake, denitrification and nitrification) are simulated using kinetic equations (Whitehead et al., 1998 a, Wade et al., 2002).

Through applying the conceptual hydrological water quality model LASCAM in the Swan-Avon catchment (area 120 000 km^2) in Western Australia, nutrient loads were found to be closely related to the distribution of rainfall and vegetation cover (Viney and Sivapalan, 2001). LASCAM uses a single set of global parameters for all the delineated sub-basins at a studied catchment. The large-scale lumped nature of the model application suggests that a complex representation was not needed. A disadvantage of LASCAM is that the model is based on the assumption that there is no significant nutrient uptake by riparian vegetation. Nitrate was concluded to be poorly predicted based on a calibration using at least a data series of 10 years of frequently collected river nutrient concentrations (Drewry et al., 2006; Viney and Sivapalan, 2001).

Through testing the INCA model in several catchments with different physiographical characteristics, it was found to be capable of reproducing the seasonal dynamics observed in the streamwater nitrogen concentration data (Wade et al., 2002, Whitehead et al., 1998 b). The internal budgeting of the model processes allows the user to investigate nitrogen dynamics and the interactions between physical, chemical and biological behaviour. The disadvantage of the INCA model is that it cannot predict the streamflow with high accuracy due to its too simplified description of hydrological processes (e.g. Wade et al., 2002, Whitehead et al., 1998 b).

SWAT has been applied worldwide for simulations of streamflow, nutrient (N and P) and sediment transport at mesoscale and large scale catchments. The SWAT model has been proved to be able to represent monthly mean streamflow and nutrient loads for long-term simulations (e.g. yearly time scale); however, the simulations of streamflow at daily time step and monthly nutrient loads (especially nitrate) seem to be still poor according to some publications (e.g. Bosch et al., 2008, Chu et al., 2004). The low performance on nitrate simulations are mainly attributed to improperly described hydrological processes (e.g. ignoring subsurface flow contributions outside the watershedand groundwater entering deep aquifers) and missing in-stream retention processes (Cherry et al., 2008; Chu et al., 2004). In addition, it requires a large number of input data for hydrological simulation and has a complex parameter set, which makes it difficult for spread application, especially for regions with a shortage of data (e.g. Chahinian *et al.*, 2011; Glavan *et al.*, 2011).

HSPF was also widely used for simulations of long-term continuous flow and nutrient loads in mixed agricultural and urban watersheds, which showed good simulation accuracy. However, similar as SWAT model, it has too many parameters required to be defined through calibration and therefore cumbersome to use (Borah and Bera, 2004; Singh *et al.*, 2005).

In hydrological water quality modeling, it is a prerequisite that a selected model has structures and parameter set of proper complexity, feeds with easily measurable data and is computationally efficient for the purposes of process understanding and catchment management. HYPE simulates discharge based on the basic and easily measurable climate data of time series precipitation and air temperature. It has a reasonable number of hydrological and water quality parameters needed to be definedin application. Through proper parameter sensitivity analysis, only a small number of sensitive parameters need to be specified through calibration. The remaining insensitive parameters can be fixed and assigned to reasonable values. There isa good balance between model complexity and representation of internal hydrological, nutrient transport and transformation processes. HYPE has been tested in several catchments of different scales (from mesoscale to national scale) in Sweden to simulate streamflow, concentrations of streamwater nitrogen and phosphorus. Good agreement between simulations and observations were reported; the temporal and spatial variations of long term average discharge and nutrient concentrations were found to be well captured by the model (Lindström*et al.*, 2010; Strömqvist *et al.*, 2012).As a newly developed hydrological model, HYPE has several similarities as the well-known SWAT model. Table 1.1 presents an overview of the two models.

Table 1.1 An overview of the hydrologic models (HYPE and SWAT).

Aspects	SWAT	HYPE
Nature	Semi-distributed hydrological model	Semi-distributed hydrological model
Scale	River basin scale, calculation at sub-basin level	River basin scale, calculation at sub-basin level
Time scale	Continuous time simulation at daily interval, not applicable of detailed, single-event flood routing	Continuous time simulation at daily interval, not applicable of detailed, single-event flood routing
Input data	Precipitation, Max/Min temperature, wind speed, relative humidity, solar radiation	Precipitation, daily mean temperature

Simulated variables	Snow melt, surface runoff and interflow, groundwater flow, reservoir routing, soil temperature, evapotranspiration, infiltration, percolation, nutrient (nitrogen and phosphorus), sediment yield, crop growth, pesticide, agricultural management	Snow melt, surface runoff and interflow, groundwater flow, reservoir routing, soil temperature, evapotranspiration, infiltration, percolation, nutrient (nitrogen and phosphorus), agricultural management
Model parameters	Related to soil type, land use and management practices	General parameters; land use, soil type or region dependent parameters

1.2.3 Applicability of hydrological water quality models

Most hydrological water quality models were developed at a specific region with certain physiographic characteristics. For example, LASCAM was developed to predict the impact of changes in climate and land use on water yield and quality (primarily salinity) in southwestern Western Australia (Viney *et al.*, 2000). HYPE model was developed based on physiographic characteristics and hydrological patterns in Sweden (Lindström*et al.*, 2010). When a hydrological water quality model is applied at one catchment, model parameters are usually calibrated against the observed responses (e.g. discharge and nutrient concentrations) from the catchment outlet. Then the model is tested in an independent period using the optimized parameters. In this way, temporal transferability of model structures and parameter set are evaluated. The model performance may not deteriorate a lot if there are no substantial differences between calibration and validation periods in terms of catchment characteristics and climate conditions (Bahremand and De Smedt, 2010).The model performance is rarely evaluated spatially, mainly due to unavailability of spatially distributed monitored data (Cherry *et al.*, 2008). When the model is applied in a new catchment, the model parameters usually need to be calibrated again due to changes of physiographic characteristics (e.g. topography, land use and soil type) and hydrological regimes.

It is important to test a hydrological water quality model in several catchments characterized by different scales, physiographic characteristics, and climate conditions. In this way, itallows for evaluating the capability of the model on representing hydrological water quality processes under different hydrological and chemical conditions. This can increase the user's confidence on the model. Despite of the inclusion of the spatially explicit representation of runoff generation processes of small landscape elements such as wetlands, the distributed hydrological model HBV was reported not able to capture major changes in runoff between different catchments using individually calibrated parameter sets according to a proxy-basin test (Wrede *et al.*, 2013). Proxy-basin tests often result in rather drastic performance reductions and are even failed by many established models (Refsgaard and Henriksen, 2004).

1.2.4 Model calibration and uncertainty analysis

1.2.4.1 Uncertainty sources

As mentioned above, nutrient leaching from landscapes to the stream involves complex hydrological transport and nutrient transformation processes. In hydrological water quality modeling, most processes are described by conceptual empirical equations based on physical understanding and knowledge of the studied catchment, physical and chemical processes, which inherently contain simplifications and assumptions. Each hydrological water quality model requires certain types of input data and has several conceptual parameters which need to be determined through calibration against observed counterparts of the simulated variables. There are four types of uncertainty sources in deterministic water flow and water quality modeling: random or systematic errors in the model inputs or boundary condition data, random or systematic errors in the recorded output data, uncertainty due to sub-optimal parameter values and errors due to incomplete or biased model structure (Butts *et al.*, 2004).

The input uncertainty contains quantities, which are temporally and spatially variable (e.g. the rainfall amount in a certain period or the soil distribution in a catchment). The uncertainty of the recorded data is related to measurement errors attributed to instability of monitoring/analysis equipment or manpower. The model parameter uncertainty is caused bythe complexity of the model structures and parameters' inter-correlation.Due to parameter interaction, different parameter sets may produce similar model outputs, which iscalled equifinality (Beven and Binley, 1992). The model structure uncertainty reflects the knowledge deficiency about the system, such as how the system is measured, understood and described. For a given model, the model structure uncertainty is expressed as parameter uncertainty, which is propagated to predictive uncertainty. Uncertainty analysis provides a methodology that can add value to conventional risk analysis by providing more information about the outputs of a predictive model, and identifies components of the model where uncertainties can be decreased. For an analyst, model user, or policy maker, uncertainty analysis has also the advantage of providing an error bound and confidence level on the output. Uncertainty analysis does not address the issue of parameter estimation itself but rather the impact of parameter uncertainty on model output uncertainty (Benke *et al.*, 2008). Therefore, it is important to estimate the predictive uncertainty caused by different uncertainty sources mainly in two aspects: (i) it helps the users to understand the limitation of the model, and (ii) it assists to provide decision makers more scientific guides for watershed management.

Input data uncertainty

The input data uncertainty is related to itsspatial and temporal resolution. For hydrological water quality models (especially distributed models (e.g. TOPMODEL)), numerous studies reported that the model scale and resolution of the digital elevation model (DEM), catchment spatial delineation (i.e. definition of hydrological response units (HRU)) affectedthe model performance (Becker and Braun, 1999; Fitz Hugh and Mackay, 2000; Wolock and Price, 1994). The dependence of model performance on the spatial scale of HRU is attributed to the various information content contained in hydrological units of different resolution. The sensitivity of simulated variables to the definition of HRU also plays an important role. Using the SWAT model for runoff and sediment predictions, FitzHugh and Mackay (2000) found that streamflow and outlet sediment predictions were not seriously affected by changes in sub-watershed size; while sediment generation estimates changed substantially between the coarsest and the finest watershed delineations due to the sensitivity of the runoff term to the size of HRUs' areas.

A hydrological model predicts runoff based on climate input data, such as precipitation, temperature, global radiation, relative humidity and wind speed. In particular, distributed rainfall-runoff models require detailed input data of high temporal and spatial resolution. However, due to technical and financial constraints, the installed net of rainfall and climate stations isin general not dense enough to cover the spatial variability of meteorological variables. As a result, the climatic data derived from insufficient meteorological monitoring stations cannot represent a real average values of the catchment/sub-basin under study due to topography effects. This situation is particularly severe for rainfall measurements, since rainfall is the most important data and it is highly variable in time and space, yet often agglomerated into a single areal average during model calibration (Kavetski *et al.*, 2006; Zawadzki, 1973). It is assumed that model calibration compromises the information loss related to less detailed input data. Andréassian *et al.* (2001) reported the reaction of watershed models to improvements in rainfall input accuracy by better performance and reduced variability of model efficiency. Faurès *et al.* (1995) also noted that insufficient knowledge of spatial rainfall variability and therefore a single rain gage station with the standard uniform rainfall assumption in a specified catchment scale can lead to large uncertainties in runoff estimation.

Model uncertainty

There are two types of model uncertainty. The first type of model uncertainty is called model structure uncertainty attributed to improper descriptions of the natural processes. Another important typeof model uncertainty is called parameter uncertainty due to model complexity and parameter inter-correlation. There were many debates, which level of model complexity is needed to capture and represent the hydrological and water quality processes at catchment scale. A model of higher complexity has more parameters.

Identification of a complex model which has a large number of parameters is problematic because (i) the limited type and resolution of observations do not have enough constraints on the simulated processes; (ii) The modeled processes may be nonlinear and the parameter inter-correlation is high. In most rainfall-runoff models, total discharge, flow components and internal hydrological variables (e.g. soil moisture and groundwater level) are simulated. However, in most cases, the hydrological parameters are calibrated only against streamflowsobserved at catchment outlet because they can be easily measured and accessed. In general, soil moisture and groundwater level measurements are not available or the observations do not have sufficient spatial and temporal resolution to constrain relevant processes and parameters. As a result, there is a high risk of over-parameterization and the optimized parameters have a high uncertainty.

The application of distributed and physically based forecasting models can provide improved streamflow forecasts. There is a trade-off between the complexity of the model descriptions necessary to capture the catchment processes, the accuracy and representativeness of the input data available for forecasting and the accuracy required to achieve a reliable, operational flood management and warning. Many studies have addressed the issues of sub-optimal parameter estimation, parameter uncertainty and model calibration. However, very few studies have examined the impact of the model structure error and complexity on the model performance and modeling uncertainty. Butts *et al.* (2004) described a general hydrological framework, which allows for the selection of different model structures within the same modeling tool. Results showed that model performance is strongly dependent on model structure and to a lesser extent on distributed rainfall; the sensitivity of the simulated streamflow to variations in an acceptable model structure are of the same magnitude as uncertainties arising from the other uncertainty sources. Wrede *et al.* (2013) reported that the distributed HBV model performed only equally well compared to the much simpler lumped HBV model with regard to simulating runoff at the catchment outlets. Refsgaard *et al.* (2006) reviewed a range of strategies for assessing structural uncertainties in models and introduced a framework for dealing with uncertainty related to model structure error. The existing strategies for model structural uncertainty analysis were categorized into "interpolation" where inferences on the accuracy of a model structure can be made directly on the basis of field data, and "extrapolation" that is beyond the situation and the field data available for calibration. For "extrapolation", it involves the use of multiple conceptual models, assessment of their pedigree and reflection on the extent to which the sampled models adequately represent the space of plausible models.

With the increasing computing power, more and more distributed hydrological and water quality models were developed. With increase of model complexity, the input data uncertainty and difficulty of parameter identification increase. Van Rompaey and Govers (2002) reported that simplification of model structure will lead to an increase of model error attributed to an incomplete description of the processes and a decrease of the input error. Therefore, for optimal model predictions at a regional scale the model complexity has to be in balance with the quality of the available input data. Perrin *et al.* (2001) implemented an extensive comparative performance assessment of the structures of 19 daily lumped hydrological models on 429 catchments and found that the complex models outperformed the simple ones in calibration mode but not in verification mode. Inadequate complexity typically results in model over-parameterization and high parameter uncertainty. Through comparing the performances of rainfall-runoff models with different complexity at three dry catchments, Gan *et al.* (1997) concluded that model performances depend more on the model structure, the objective function used in automatic calibration, and data quality, than on model complexity or length of calibration data. Refsgaard and Knudsen (1996) compared three models (a lumped conceptual modeling system (NAM), a distributed physically based system (MIKE-SHE), and an intermediate approach (WATBAL)) for runoff simulation at three catchments. They reported that all three models performed equally well when at least a data series of one year was available for calibration, while the distributed models performed slightly better for cases where calibration was not allowed.

There were also several studies conducted in the past decades to investigate the effects of water quality model complexity on model performances, parameter identification and predictive capability. De Wit and Pebesma (2001) applied four models of different complexity in two large European river basins (Rhine and Elbe) to simulate the transfer of nitrogen and phosphorus from pollution sources to river outlets. It was found that the addition of more processes' description does not necessarily improve the predictive capacity because the information content of the available database is only sufficient to support models of limited complexity. Van der Perk (1997) compared 8 one-dimensional steady-state models with different complexity for simulating the riverflow phosphate concentration and noted that the identifiability of the model parameters becomes poorer with increasing model complexity.Model predictive accuracy decreases with increasing model complexity (i.e. the uncertainty in model results become larger)if the correlation between the model parameters is not taken into account.

All model structures must, to some extent, be in error and all observations and measurements on which model calibration is based must also be subject to error. Therefore, it is not reasonable to expect that any set of parameter values will represent a true parameter set to be found by some calibration procedures. It is difficult to find an optimal parameter set in the high-dimensional parameter space associated with hydrological water quality models due to the use of the threshold parameters, inter-correlation between parameters, autocorrelation and heteroscedascity in the residuals and insensitive parameters. As mentioned in Beven and Binley (1992), even when the hydrologist is prepared to accept that a distributed model is predicting the right sort of response mechanism, there may be many different combinations of the grid element parameters that might lead to equivalently accurate predictions (i.e. equifinality). Moreover, it is easy to show that if the same model is "optimized" on two different periods of record, two different optimal parameter sets will be produced (Beven, 1993). Given the complexities of a watershed and the large number of interactive processes occurring simultaneously (such as with SWAT model), a hydrological water quality model calibrated based on measured data from the watershed outlet may produce erroneous results for various land uses and sub-basins within the watershed, unless the objective function was constrained to produce correct results (Abbaspour et al., 2007). Thus, it is important to assess the parameter and predictive uncertainty associated with the best estimates.

1.2.4.2 Model calibration and uncertainty analysis methods

<u>Model calibration approaches</u>

Hydrological water quality model calibration is a process that optimizes the parameters' values by minimizing the distance between the model-simulated outputs and the observed data. The distance is defined by an objective function based on certain criteria. There are two types of calibration approaches: manual calibration and automatic calibration. Manual calibration is implemented through iterative trial and error based on experience and knowledge of the studied catchment and hydrological water quality processes. With manual calibration, parameters are adjusted one by one. Parameters' inter-correlation cannot be considered, which may result in the problem of over-parameterization. Automatic calibration optimizes the model parameters by minimizing the objective function using certain algorithm. The automatic calibration is able to take advantage of the speed and power of digital computers, while being objective and relatively easy to implement. Boyle et al. (2000) suggested combining the strengths of manual and automatic calibration methods to improve optimization of hydrologic models in the way that the attention of the hydrologist can be redirected from the tedious effort of manually searching for the "good" region to the more productive task of evaluating solutions from within the region found using the automatic search algorithm. In both manual and automatic calibration methods a historical period of input and output data records specific to the site in which the model is being applied should be selected in a way that the data are considered to be fairly well representative of the various phenomena that the system experiences and that the model is intended to simulate (Sorooshian and Gupta, 1983).

There are several automatic parameter optimization approaches, which are sorted into local search approaches and global search approaches. Local search approach employs the hill climbing strategy, which can be further divided into "direct" and "gradient-based" methods. Direct search methods only use information on the objective function value, whereas gradient-based methods also use information about the gradient of the objective function. Local search methods are efficient for locating the optimum of a unimodal function since in this case the hill-climbing search will eventually reach the global optimum, irrespective of the starting point (Madsen, 2000). One of most popular direct search approach is the simplex method (Nelder and Mead, 1965). Gradient-based methods include the steepest descent method and various approximations of the Newton method (e.g. the Gauss-Marquardt algorithm). PEST (Model-Independent Parameter Estimation & Uncertainty Analysis) is an automatic local search tool for parameter optimization and predictive analysis that uses the Gauss-Marquardt-Levenberg algorithm (Doherty, 2005).PEST uses the hill climbing technique, following the steepest gradient of the objective function, starting at a specified point in the parameter space (Rode *et al.*, 2007 b). It has been widely used for parameter optimization and predictive analysis of hydrological and water quality models (e.g. Bahremand and De Smedt, 2010; Rode *et al.*, 2007 b).

Conceptual rainfall-runoff models may have numerous local optima on the objective function. In such cases local search methods are inappropriate because the estimated optimum will depend on the starting point of the search (Duan, 1992). For such multi-modal objective functions, global search approaches which denote those algorithms designed for locating the global optimum and not being trapped in local optima should be applied. Shuffled Complex Evolution (SCE-UA) global optimization algorithm has been widely used and proved to be consistent, effective, and efficient in locating the global optimal parameter values of hydrological models (e.g. Duan *et al.*, 1992; Van Griensven and Bauwens, 2003). Vrugt *et al.* (2003) presented the Multi-objective Shuffled Complex Evolution Metropolis (MOSCEM) algorithm, which uses the concept of Pareto dominance rather than direct single-objective function evaluation to evolve the initial population of points towards a set of solutions stemming from a stable distribution (Pareto set). Through comparing three optimization methods, namely SCE-UA, the Multiple Start Simplex (MSX), and the local Simplex on 32 CRR-catchment case studies (combination from four rainfall-runoff models (CRR) and eight catchments), Gan and Biftu (1996) found that both SCE-UA and the local Simplex are viable optimization tools, while MSX is computationally insufficient. SCE-UA can complete the parameter search in one run, while the local Simplex often requires multi-run operations to get good results.

Uncertainty analysis approaches

In past years several algorithms have been developed for implementing uncertainty analysis on hydrological models. Some examples are the Generalized Likelihood Uncertainty Estimation Methodology (GLUE) (Beven and Binley, 1992), Multi-Objective Complex Evolution (MOCOM-UA) (Yapo *et al.*, 1998), Bayesian Recursive Estimation (BaRE) (Thiemann *et al.*, 2001), PEST (Doherty, 2005), Shuffled Complex Evolution Metropolis algorithm (SCEM-UA) (Vrugt *et al.*, 2003) and DREAM$_{(ZS)}$ (DiffeRential

Evolution Adaptive Metropolis algorithm) (Laloy and Vrugt, 2012). Uncertainty analysis is usually carried out during model calibration process. Therefore, these algorithms were also used for model calibration. Similar as category of calibration approaches, uncertainty analysis algorithms are sorted into local approaches (e.g.PEST) and global approaches (e.g. GLUE, MOCOM-UA, BaRE, SCEM-UA and DREAM$_{(ZS)}$). GLUE enables to incorporate different types of observations into the calibration, Bayesian update of likelihood values and evaluates the value of additional observations to the calibration process. In the application of GLUE, it involves five procedures: (i) A formal definition of a likelihood measure or set of likelihood measures. The choice of a likelihood measure is inherently subjective. (ii) An appropriate definition of the initial range or distribution of parameter values to be considered for a particular structure. There may be a considerable degree of subjectivity at this point due to lack of prior knowledge about the parameter values. (iii) A procedure for using likelihood weights in uncertainty estimation. (iv) A procedure for updating likelihood weights as new data becomes available. In the procedure of updating likelihood weights using Bayes equation, the definition of the distributions remains subject to the sampling limitations of the Monte Carlo procedure. (v) A procedure for evaluating uncertainty such that the value of additional data can be assessed.

The theory of applying PEST for predictive uncertainty analysis is described briefly as follows: if the minimum of the objective function is determined as during calibration process and if all parameters for which the objective function is less than (where δ is relatively small) can also be considered to calibrate the model, then the range of parameter values which can be considered to calibrate the model can be quite large indeed. To fully explore the repercussions of parameter non-uniqueness on predictive uncertainty, parameters must be varied in such a way that the objective function hardly changes which can be realized by defining δ properly. After thousands of model runs have been undertaken a suite of predictions will have been built up, all generated by parameter sets which satisfy calibration constraints. In this way, the predictive uncertainty induced by parameter non-uniqueness can be investigated. Bahremand and De Smedt (2010) combined PEST with the hydrologic model WetSpa at a rather large catchment (1297 km^2) for parameter estimation, sensitivity and predictive analysis. They found that the correction factor for measured evaporation data has the highest relative sensitivity and the parameter uncertainty does not result in a significant level of predictive uncertainty. Doherty and Johnston (2003) presented the way to apply regularization and nonlinear predictive uncertainty analysis functionality of PEST to investigate the parameter non-uniqueness and their effects on model predictive uncertainty through a case study. Rode *et al.* (2007 a) investigated the information contents of different calibration data sets using PEST. They noted that parameters' uncertainty decreased and model validation performance improved with increasing number of data sets included in the calibration procedure and declared that these findings are restricted to cases where data sets of different conditions are available.

SCEM-UA is an effective and efficient evolutionary sampler, which is a modified version of the original SCE-UA global optimization algorithm developed by Duan *et al.* (1992). It merges the strengths of the Metropolis algorithm, controlled random search, competitive evolution, and complex shuffling in order to continuously update the proposal distribution and evolve the sampler to the posterior target distribution. Compared to traditional MCMC (Markov Chain Monte Carlo) samplers, SCEM-UA algorithm is an adaptive sampler, where the covariance of the proposal or sampling distribution is periodically updated in each complex during the evolution to the posterior target distribution using information from the sampling history induced in the transitions of the generated sequences. The SCEM-UA algorithm is different from the original SCE-UA algorithm in two aspects. Firstly, the downhill Simplex method in the competitive complex evolution algorithm outlined by Duan *et al.* (1992) is replaced by a Metropolis annealing covariance based on offspring approach, thereby avoiding a deterministic drift toward a single mode. Secondly, the SCEM-UA algorithm does not further subdivide the complex into sub-complexes during the generation of the offspring (candidate points) and uses a different replacement procedure, to counter any tendency of the search to terminate occupations in the lower posterior density region of the parameter space. Both modifications are necessary to prevent the search from becoming trapped in a small parameter space of attraction and therefore to arrive at the correct posterior target distribution.

Through comparison of MH (Metropolis-Hastings algorithm) with SCEM-UA in implementing uncertainty analysis of a five parameter conceptual rainfall-runoff model HYMOD, it was found that the SCEM-UA algorithm is more efficient in traversing the parameter space, with convergence to a stationary posterior distribution than the traditional MH sampler (Vrugt *et al.*, 2003). This was explained by the ability of the SCEM-UA algorithm to exchange information about the search space gained by the different parallel launched sequences, which increases the explorative capabilities of the sampler and therefore the traversing speed of the chains through the feasible parameter space. Periodic shuffling of the complexes in the SCEM-UA algorithm ensures sharing of information gained independently by each community about the nature of the posterior distribution and thus increases the traversal through the parameter space. $DREAM_{(ZS)}$ capitalizes on the advantages of DREAM (differential evolution adaptive Metropolis) for posterior exploration but generates candidate points in each individual Markov chain by sampling from an archive of past states. $DREAM_{(ZS)}$ algorithm is able to execute N different candidate points simultaneously in parallel (Laloy and Vrugt, 2012).

Mousavi et al. (2011) compared different calibration procedures (i.e. scenarios of separately and jointly-calibrated events) on parameter identification by linking the SUFI2 (Sequential uncertainty fitting) technique with HEC-HMS (Hydrologic Engineering Center-Hydrologic Modelling System) hydrologic model and testing it in Tamar basin located in south of Iran. They found that the SUFI2 technique linked to HEC-HMS as a simulation-optimization model can provide a basis for performing uncertainty-based automatic calibration of event-based hydrologic models. Gardner et al. (2011) combined Bayesian MCMC with the Big Sky nutrient model (BiSN) for model specification, which revealed successfully model and parameter uncertainty as well as the primary processes governing watershed NO_3^- export in watershed areas with short travel times to the stream. Zheng et al. (2011) applied Probabilistic Collocation Method (PCM) to a WARMF model for simulating diazinon pollution to assess the modeling uncertainty. They reported that the PCM-based approach is more efficient than conventional Monte Carlo methods regarding computational time as well as providing insights into data collection, model structure improvement and management practices. Based on the review of former uncertainty analysis studies on hydrological and water quality modeling, MCMC algorithms are concluded to be more efficient and objective than other approaches to evolve parameter posterior distributions from the initial whole parameter ranges and estimate predictive uncertainty.

1.2.5 Step-wise calibration vs. multi-objective calibration

In hydrological water quality modeling, step-wise calibration is traditionally used for parameter optimization under which the hydrological parameters are calibrated first and then the water quality parameters are calibrated (e.g. Arheimer and Brandt 1998; Strömqvist et al., 2012). With step-wise calibration, it is risky that water quality is calibrated, subsequently, to a poor calibration of the flow (van Griensven and Bauwens, 2003). Numerous studies have reported that multi-objective calibration (here it means calibrating hydrological and water quality parameters simultaneously) using corresponding observations (e.g. discharge and nutrient concentrations) is more appropriate for parameter optimization in integrated hydrological and water quality modeling (Lu et al., 2011; Rode et al., 2007 b, van Griensven and Bauwens, 2003). Multi-objective calibration increases constraints on both hydrological and water quality processes as they are correlated (Basu et al., 2010 and 2011; Lam et al., 2012; Li et al., 2010; Liu et al., 2008; Rode et al., 2008; Shrestha et al., 2012; Wriedt and Rode, 2006). Also, adding water quality observations into hydrological parameters identification helps to inform partitioning the river flow into surface flow, interflow, and groundwater flow components, as flow contributions differ in the types and concentration levels of the pollutants they carry (Sivapalan et al., 2003). Through calibrating hydrological and water quality parameters simultaneously by minimizing a properly built objective function, the information content of different types of observations can be fully exploited. Therefore, it may improve the identification of both hydrological and water quality parameters. Automatic multi-objective global optimization algorithm based on SCE-UA has been developed and employed for calibration of the semi-distributed hydrological water quality model SWAT (Van Griensven and Bauwens, 2003).

1.2.6 Effects of spatial and temporal resolution of calibration data on model identification

Effects of spatial resolution of calibration data on model identification

In general, hydrological and water quality parameters are calibrated against discharge and concentrations of water quality constitutes (e.g. N and P) measured at catchment outlet (i.e. single-site calibration). This is because hydrological water quality monitoring/sampling is often implemented only at catchment outlet due to technical or financial constraints. This approach is appropriate to some extent as the streamflow and nutrient concentrations observed at catchment outlet represent the overall response of the whole catchment to the climate forces and land management practices. However, the parameters optimized in this way may contain high uncertainty if a distributed/semi-distributed hydrological water quality model is applied and the studied catchment shows high spatial heterogeneity in terms of topography, soil type, land use and underlying geology. The sensitivity of land use- or soil type- dependent parameters are highly dependent on the dominance of certain land use or soil type (Feyen and Vázquez, 2000; van Griensven and Bauwens, 2003; Zhang *et al.*, 2008). Moreover, the hydrological/water quality parameter set optimized through calibration against relevant observations from catchment outlet may not predict discharge/nutrient concentrations accurately in internal sites (Moussa*et al.*, 2007; Refsgaard, 1997; Zhang *et al.*, 2008). Therefore, hydrological and water quality observations from catchment internal sites may be needed to increase constraints onthe related processes if the catchment is spatially heterogeneous. However, very few studies have been reported on the effects of spatial resolution of discharge/nutrient observations on identification of hydrological water quality models and predictive accuracy.

As mentioned before, single-site calibration and split-sample test are commonly used for hydrological water quality parameter optimization and model verification. Split-sample test means calibrating a model using observations of one period and validating the model in an independent period with the optimized parameter set. Refsgaard (1997) reported that model prediction accuracy decreased significantly when the optimized parameter values from single-site calibration were tested in internal sub-catchments. Therefore, it was suggested that multi-site calibration/validation is needed if spatially distributed predictions are required. With the development of hydrologic monitoring technique, more and more spatially distributed observatory systems are installed and multi-site measurements are available. In recent years multi-site multi-variable calibration and validation approaches were applied in hydrologic water quality simulations (Cao *et al.*, 2006; Moussa *et al.*, 2007; Refsgaard, 1997; van Griensven and Bauwens, 2003; Wang *et al.*,

2012; Zhang et al., 2008). Through comparing predictive capacity of the spatially distributed hydrologic model ModSpa using parameter sets calibrated against discharge measurements from different number of gauging stations, Moussa et al. (2007) found that model performance improved gradually as more flow measurements became available. Zhang et.al (2008) reported that parameter values estimated through simultaneous multi-site calibration outperformed those calibrated against discharge observations from a single monitoring site, which indicates the importance of spatially distributed data in hydrologic parameter identification, especially in heterogeneous catchments. However, Wang et al. (2012) concluded that multi-site calibration protocol did not improve model simulation results compared with single-site calibration protocol through a case study using MIKE-SHE. It was commented that parameters are usually better identified when new types of field data are used for calibration rather than adding more data of the same variable (Madsen, 2003). Thus, there are no consistent conclusions on the effects of multi-site calibration strategy on parameter identification, model performance and prediction accuracy.

Effects of temporal resolution of calibration data on model identification

Water quality samplings (e.g. N and P concentrations) are in general sparser compared with discharge observations in terms of temporal resolution due to financial and personal constraints (Ullrich and Volk, 2010). For instance, nutrient concentrations are usually sampled and analyzedat weekly to monthlytime interval. In most hydrologic water quality modeling studies, water quality parameters were calibrated against nutrient concentrations/loads observed/calculated using non-continuous observations at biweek-ly/monthly time step (e.g. Lindströmet al., 2010; Nafees Ahmadet al., 2011). There are mainly two risks related to the water quality parameter identificationusing this type of sparse measurements. (i) The temporal variation of nutrient concentrations may be underestimated with biweekly/monthly water quality samplings due to variations of nutrient inputs and flow conditions (Basu et al., 2011; Onderka et al., 2012; Li et al., 2010; van Griensven et al., 2006), which may result in high uncertainty in the optimized water quality parameter values. (ii) It may result in high uncertainty in nutrient load predictionusing the optimized water quality parameters calibrated against non-continuous sparse measurements, especially if peak flow events are not covered in the sampling campaigns (Li et al., 2010; van Griensven et al., 2006, Shrestha et al., 2012).

It is important to estimate nutrientleaching load at various time intervals (daily to yearly) for watershed management. Using regular nitrogen concentration measurements (weekly to monthly) and continuous streamflow observations, interpolation/extrapolation approach is often applied to estimate nutrient load over longer periods based on relationship between nitrogen concentration and streamflow. However, this may result in errors in load estimation due to ignorance of temporal variation of nitrogen concentrations during short period and improper assumptions used in interpolation/extrapolation approaches. It has been reported that the accuracy of load estimation depends on the sampling method and frequency, load estimation methodology and the period concerned (Littlewood, 1995; Rode and Suhr, 2007a; Rode *et al.*, 2007b; Ullrich and Volk, 2010). Because of spatial and temporal variability of hydrological regimes,nutrient transport and transformation related to climate variation and catchment heterogeneity, it is important to implementan appropriate streamflow and nutrient monitoring campaignof sufficient spatial and temporal resolution for identifyingmodel parameters, decreasing parameter- and predictive- uncertainty. Uncertainty analysis allows for estimating the parameter- and predictive- uncertainty as well as assessing the information content ofdifferent monitoring networks, which can in turn provide useful guidelines for optimal hydrological water quality monitoring.

Through simulating streamflow, nitrate load, sediment load and implementing uncertainty analysis in the pre-alpine/alpine Thur watershed using SWAT model, Abbaspour *et al.* (2007) concluded that a watershed model calibrated based on measured data at the outlet of the watershed may produce erroneous results for various landuse and sub-basins within the watershed, unless the objective function was constrained to produce correct results, which indicates that large amount of measured data are necessary for a proper model calibration. Beven (2007) pointed out the importance to study the value of different types of data in terms ofcontinued monitoring and guided field campaigns, learning about places and constraining predictive uncertainties. Rode and Suhr (2007b) noted that water quality sampling location has considerable effect on the representativeness of a water sample and the sampling uncertainties are highly site specific, therefore, spatial variation of concentrations within a cross section can make a substantial contribution to the uncertainty of transport estimates; different water quality constitutes may require different sampling frequency for reliable annual load estimates related to respective magnitude of temporal variation.

1.3 Knowledge gaps

According tothe above intensive literature review, there are several hydrological water quality models being capable of simulating streamflow and nutrient transport at catchment scale.Some of the models have too complex structures and parameter setsand require too many types of input datathat are not always available (e.g. HSPF, SWAT). Some of the models cannot capture hydrographs/nitrogen concentration dynamics satisfactorily due to improper descriptions of hydrological/nitrogen processes (e.g. INCA, SWAT and LASCAM). Therefore, it is necessary to introduce a hydrological water quality model that has model structures and a parameter set of reasonable complexity and feeds with easily measurable input datato be used as a watershed management tool. The HYPE model satisfies these requirements and has been verified to simulate streamflow and nutrient dynamics at different catchment scales in Sweden

(Lindström et al., 2010; Strömqvist et al., 2012). However, the HYPE model has not been tested in regions that have different climate and physiographic characteristics compared with Sweden. The applicability of HYPE modelacross different climatic, hydrological and biochemical conditions needs to be further tested.

As mentioned in section 1.2.5, a step-wise calibration is often usedin integrated hydrological water quality modeling. Although there were some comments that multi-objective calibration (calibrating hydrological and water quality parameters simultaneously) allows for increasing constrains on both processes (e.g. Lu et al., 2011; van Griensven and Bauwens, 2003). According to the author's literature review, no study was implemented to compare step-wise calibration and multi-objective calibration quantitatively in terms of parameter sensitivity, optimized parameter values, model performance and robustness.

In uncertainty analysis of hydrological modeling, most studies were focused on the investigation of input data uncertainty, model parameter uncertainty and their effects on predictive uncertainty (e.g. Bahremand and De Smedt, 2010; Vrugt et al., 2003). There were also some studies on the effects of model complexity on model predictive accuracy and robustness (e.g. Perrin et al.2001; Van Rompaey and Govers 2002). Just a few studies investigated the influence of spatial and temporal resolution of calibration data (discharge and nitrogen concentration) on the parameter identification, model performance and predictive accuracy.Very few studies have been carried outon uncertainty analysis for nitrogen modeling due to complexity of nitrogen transport processes and coarse temporal resolution of observed nitrogen concentrations.

In hydrological water quality modeling, model parameters can be optimized through manual calibration, automatic calibration using local search algorithm (e.g. PEST) or global search algorithm (e.g. $DREAM_{(ZS)}$). Manual calibration is tedious and time consuming, but supports the user to gradually get familiar with the catchment characteristics and model performance. PEST is computationally efficient and was proved to be effective in calibrating hydrological water quality models (e.g. Bahremand and De Smedt, 2010; Rode et al., 2007 b). But the parameter's initial values and ranges need to be defined properly in order to obtain reasonable optimal values through a small number of model runs. $DREAM_{(ZS)}$ is more computationally cost expensive. However, only proper parameter ranges need to be defined during calibration processes. $DREAM_{(ZS)}$ allows for searchingthe whole parameter spaces for the optimal parameter values, evolving the parameter posterior distributions and estimating the predictive uncertainty objectively. Concerning automatic calibration, no study was reported on comparing local search and global search approaches in terms of effectiveness, efficiency, model performance and robustness of the optimized parameter values, and capacity of predictive analysis.

1.4 Objectives

Based on above literature review and summarized knowledge gaps, this study has five objectives:

(i) Test the applicability of HYPE model for simulating streamflow and streamwater IN concentration at two meso scale catchments in central Germany (Selke and Weida) characterized by different hydrological and chemical patterns. In addition, these two catchments show different climatic and physiographic characteristics compared with Sweden where the HYPE model was developed and intensively applied.

(ii) Evaluate two different model calibration strategies (step-wise calibration and multi-objective calibration) on hydrological water quality parameter optimizationin terms of model performance and robustness of the optimized values. This will be realized by comparing parameter calibration results derived from both calibration strategies at Selke and Weida catchments.

(iii) Assess the effects of spatial resolution of calibration data (discharge and streamwater IN concentration observations) on model identification. The effects of spatial resolution of calibration data on model identification will be investigated by comparing parameter posterior distributions, model performance and robustness of the optimized parameter set, predictive uncertainty ranges derived from single-site and multi-site calibrations.

(iv) Assess the effects of temporal resolution of calibration data (IN concentration observations) on model identification. The effects of temporal resolution of calibration data (streamwater IN concentration observations) on model identification will be investigated by comparing parameter posterior distributions, model performance and robustness of the optimized parameter set, predictive uncertainty ranges and load prediction accuracy derived from calibration and validation using biweekly nitrate-N concentration measurements and daily averages of nitrate-N concentration observations. To this end, both local search approach (PEST) and global search approach (DREAM$_{(ZS)}$) will be combined with the HYPE model to implement parameter automatic calibration and predictive analysis in Selke catchment.

(v) Evaluate the model identification results derived from calibrations using PEST and DREAM$_{(ZS)}$ in terms of computational efficiency, parameter posterior uncertainty ranges, model performance and robustness of the optimized values, and predictive uncertainty. Through overall comparison between PEST and DREAM$_{(ZS)}$, advantages and shortcomings of both approaches will be investigated. Finally, a proper choice of calibration and uncertainty analysis approach will be suggested based on different objectives.

1.5 Structure of Dissertation

This dissertation consists of sixchapters. A brief description of each chapter is given below.

In Chapter 1, an overall introduction of this study was given. Firstly, the background of this study (problem of freshwater eutrophication) was explained. Then an intensive literature review (state of art) on hydrological water quality processes at catchment scale, modeling approaches, model applicability, model uncertainty analysis, calibration strategies for optimization of integrated hydrological water quality models, and on effects of spatial and temporal resolution of calibration data on model identification was presented. Thirdly, deficiencies and knowledge gaps in hydrological water quality modeling studies related to the issues mentioned above were declared. Based on the literature review and summary of knowledge gaps, the objectives of this study were derived. At last, a brief introduction of the structure of the whole dissertation was given.

Chapter 2 presented firstly the study catchments (Selke and Weida). Then the hydrological water quality model HYPE was described. Thirdly, methodologies of parameter automatic calibration and predictive/uncertainty analysis (PEST and DREAM$_{(ZS)}$) were introduced. Fourthly, the HYPE model setup at Selke and Weida catchments, procedures of multi-site and multi-objective calibration, procedures of stepwise calibration and multi-objective calibration, and the procedures model identification using different temporal resolution of IN concentration observations.

In Chapter 3, streamflow and IN simulation results at Selke catchment were presented and discussed. This study has been published in the journal Ecohydrology (Jiang *et al.* (2014)). Hydrological and water quality parameters were optimized using multi-site and multi-objective calibration with PEST. Sensitive hydrological nitrogen transport processes and relevant parameters were discussed based on sensitivity analysis. Model performances on simulations of discharge and soil moisture, streamwater IN concentration and IN load were shown and discussed. Temporal and spatial variations of runoff and streamwater IN concentrations related to climate variability and catchment heterogeneity were investigated.Based on analysis of the temporal and spatial variations of IN load, the controlling effect of discharge on IN leaching at this catchment was determined.

Chapter 4 consists of two parts. In the first part, HYPE model was set up at Weida catchment for simulating streamflow, streamwater IN concentration and load that shows different catchment characteristics, hydrological and chemical patterns compared with Selke. In this way, model applicability at catchments with various physiographical characteristics was further assessed. In the second part, two different calibration strategies, namely step-wise calibration and multi-objective calibration, were evaluated through case studies at Selke and Weida catchments. Parameter identification results, model performance and robustness of the optimized parameters derived from these two different calibration strategies were compared.

Accordingly, an optimal calibration procedure for parameter optimization in integrated hydrological water quality modelling was suggested.

In Chapter 5, the effects of spatial and temporal resolution of calibration data (discharge and streamwater IN concentration observations) on model identification were evaluated with focus on the Selke catchment. PEST and DREAM$_{(ZS)}$ were combined with the HYPE model to implement an automatic parameter calibration and uncertainty analysis. The impacts of spatial resolution of daily mean discharge and streamwater IN concentration measurements on model identification were investigated by comparing model calibration and validation results, parameter posterior distributions and prediction uncertainty derived from single-site calibration and multi-site calibration. The effects of temporal resolution of calibration data (streamwater IN concentration measurements) on model identification were investigated by comparing model performance, parameter posterior distributions, prediction uncertainty and IN load prediction accuracy derived from calibration and validation using biweekly nitrate-N concentration measurements and daily averages of nitrate-N concentration observations. Parameter identification results obtained from PEST and DREAM$_{(ZS)}$ were compared regarding effectiveness and efficiency, capacity to evolve parameter posterior distributions and estimate predictive uncertainty. Accordingly, appropriate approach was suggested for parameter identification in integrated hydrological water quality modeling based on different purposes.

In Chapter 6, conclusions of this study were given.

Chapter 2: Materials and Methodologies

2.1 Selke catchment

The study catchment (Selke) is a tributary of the Bode River. It is a mesoscale, lower-mountain range catchment covering an area of 463 km^2 at the outlet gauging station Hausneindorf (Figure 2.1a). The Selke river basin has three gauging stations (Silberhuette, Meisdorf and Hausneindorf), where discharge and IN concentrations were measured. The Selke River originates from the Harz mountain range and discharges into the Bode River in the lowland areas. From headwater to catchment outlet, the elevation varies from 605 to 53 m (Figure 2.1a). Land use is dominated by forest (such as broad-leaved forest, coniferous forest and mixed forest) in the mountain areas and agriculture in the lowland areas (Figure 2.1b). The shares of agriculture and forest in the Selke catchment are 52% and 35%, respectively. From upstream to downstream, a decrease of forest and an increase of agriculture land can be observed. Soil is dominated by cambisols in the mountain areas and chernozems in the lowland areas (Figure 2.1c). The underlying geology is characterized by schist and claystone in the upstream areas and tertiary sediments with loess in the downstream areas.

The mean annual precipitation decreases from 792 mm in the Harz Mountain to 450 mm in the lowland areas, with an average of 660 mm for the whole Selke catchment (Haberlandt *et al.*, 2008). More precipitation falls in summer, with a ratio between summer and winter of 1.35. The mean temperature is 9°C, with an average monthly low of - 1.8°C in January and high of 15.5°C in July. There is an increase of temperature from mountain areas to the downstream areas due to the elevation effect. Nutrient inputs from agricultural land are the main eutrophication sources of the streamwater. Main crops include winter wheat, triticale, winter barley, rye, rape and corn. Additionally, sugar beet is grown in the fertile lowland areas of the catchment. Kistner (2007) has reported that the fertilizer inputs range from 130-190 kg N ha^{-1} y^{-1} to 20-30 kg P ha^{-1} y^{-1}, referring to a survey with farmers in the Selke catchment. The long-term mean discharge is 1.54 m^3 s^{-1}. The mean IN concentration measured at the stations Silberhuette, Meisdorf and Hausneindorf is 1.44, 1.75 and 3.91 mg l^{-1}, respectively. It has to be mentioned that stream water was pumped for the flooding of two mining lakes in the lowland areas during the period of Nov. 1998-Dec. 2004, with an average pumping rate of 1000 m^3 h^{-1}. The locations of water abstraction and lakes are shown in Figure 2.1a and Figure 2.1b, respectively. Streamflow shows substantial temporal variations, which is characterized by high flow during the winter (due to significant snowmelt in addition to the rainfall) and low flow with occasional peak flows caused by high rainfall events in summer. The summaries of catchment characteristics for the whole Selke catchment and sub-basins Silberhuette and Meisdorf are given in Table 2.1.

There are 16 precipitation stations and 2 climate stations within/close to the Selke catchment. Precipitation stations are denser in the mountain areas compared with the lowland areas. Availability and resolution of spatial and time series data used for model setup are shown in Table 2.2. Calibration and validation data of daily mean discharge have been measured at gauging stations Silberhuette, Meisdorf and Hausneindorf (Figure 2.1a). Considering the water abstraction from Selke River to the lakes during 1999-2004 (Figure 2.1a), the average yearly extraction amount and change in lake volume are 3.22 million m^3 and 5.53 million m^3, respectively. The difference between extraction and change in lake volume is explained by the additional groundwater recharge induced by streamwater abstraction. The streamwater nitrogen concentrations were sampled at the same location as the discharge only for the station Silberhuette. For the stations Meisdorf and Hausneindorf, the water quality samples were taken 2-4 km downstream from the corresponding discharge gauging station. The sampling frequency varied from weekly to monthly time steps. While, the station Hausneindorf (catchment outlet) had more frequent nitrogen sampling compared with the other two stations. For nitrogen, concentrations of nitrate-nitrogen, ammonium-nitrogen, nitrite-nitrogen and mineral-nitrogen were determined. The calibration and validation data of IN concentration equals mineral-nitrogen, which is approximately the sum of nitrate-nitrogen, ammonium-nitrogen and nitrite-nitrogen. The time series data of discharge and IN concentrations observed during the period 1994-2004 were used in this study for model calibration and validation.

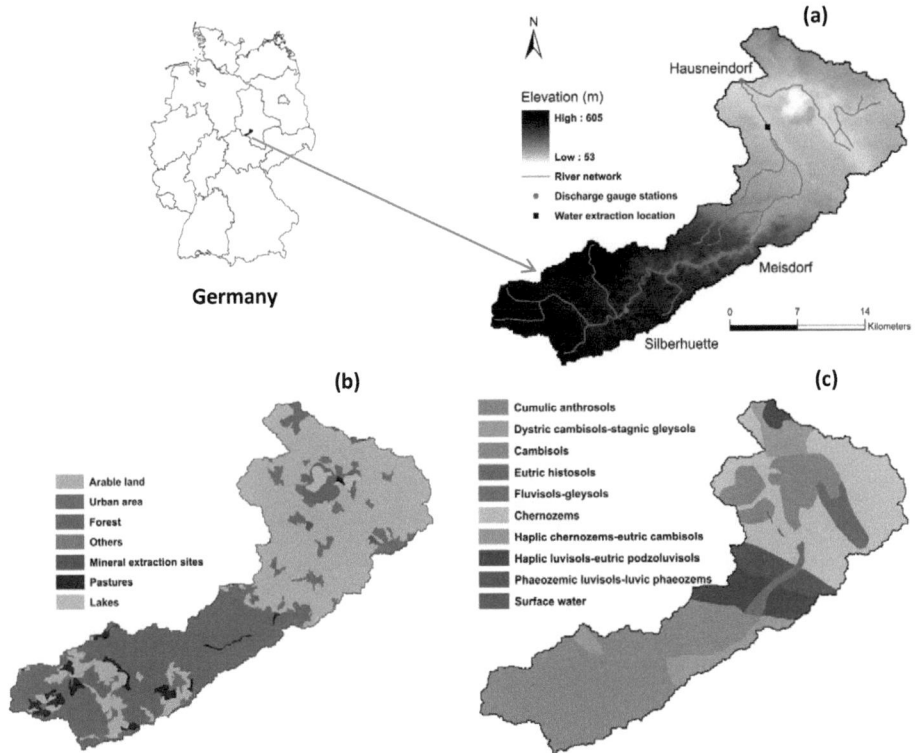

Figure 2.1 The geographical location of the Selke catchment and its DEM are illustrated in panel (a). Characteristics of land use and soil are shown in panels (b) and (c), respectively.

Table 2.1 Catchment characteristics, mean specific discharges and mean IN concentrations of the whole Selke catchment and its internal sub-basins Silberhuette and Meisdorf (Figure 2.1). The mean annual precipitation and the mean specific discharge of each concerned catchment, and mean IN concentration observed from each gauging station were calculated based on respective 10-year measurements (1994-2004).

	Selke	Silberhuette	Meisdorf
Mean elevation range (m)	104-469	409-469	212-469
Area (km^2)	438	98.7	176.3
Geology	Schist and claystone in the mountain area, tertiary sediments with loess soil in the lowland areas	Schist and claystone	Schist and claystone
Dominant vegetation type	Agriculture land and forest	Forest	Forest
Share of forest (%)	35.4	60.4	71.9
Share of arable land (%)	52.3	25.3	16.9
Mean annual precipitation (mm y^{-1})	Mountain areas: 792 Lowland areas: 450	642	658
Mean specific discharge (l s^{-1} km^{-2})	3.99	13.15	8.40
Mean IN concentration (mg l^{-1})	3.91	1.44	1.75

Table 2.2 Description of spatial and time series input data for the HYPE model setup in the Selke catchment.

Data type	Data description/properties	Resolution	Source
Geographical data	Elevation	90 m	State Survey Office
	Stream network	-	State Survey Office
	Soil type	50 m	State Survey Office
	Land use	25 m	Corine Land Cover 2006
Meteorological data	Daily precipitation and mean air temperature	16 rainfall and 2 climate stations	German Weather Service-DWD
Agricultural practices	Manure and inorganic fertiliser application, timing and amount for fertilisation, sowing and harvesting	-	Field survey
Soil nitrogen content	Initial nitrogen storage	-	Literature review
Sewage treatment plants	Water flow and IN concentration	Constant daily loadings from 6 sewage treatment plants	Operating reports of sewage treatment plants

2.2 Weida catchment

The Weida stream is a small tributary of the Weiße Elster river in the Elbe river basin, Germany (Figure 2.2 a). The Weida catchment has a size of 99.5 km^2 with the outlet Laewitz discharge gauging station and is located in the Thuringian State Mountains. The elevation of Weida catchment varies between 357 m in the northern part and 552 m in the southern part (Figure 2.2 a). The geology is dominated by clay schist and eruptive rocks. Most of these rocks have low permeability. The soils developed from this bedrock range from shallow rankers to well-developed cambisols and fluvisols in the stream valley. Dominant soil classes are sandy loam (40%) and silt loam (36%) (Hesser et al., 2010; Shrestha et al., 2007). The main types of land use include arable land (40%), forest (29%) and grassland (26%). The maps of land use and soil type are presented in Figures 2.2 b and 2.2 c, respectively. Agricultural land in Weida catchment is moderately intensive. Arable crops are dominated by winter wheat (Triticum aestivum L.), winter barley (Hordeum vulgare L.), rape (Brassica napus L.), and maize (Zea mays L.). Sugar beets (Beta vulgris L.), potato (Solanum tuberosum L.), and other intensive crops have only a small part in the cropping systems (about 3%). Fertilizer application amounts vary between 125 kg ha^{-1} y^{-1} (winter wheat) and 150 kg ha^{-1} y^{-1} (maize) for winter grain and maize and are lower for summer grain, such as summer barley (70 kg ha^{-1} y^{-1}) and oats (90 kg ha^{-1} y^{-1}). Organic farming was conducted on <2%

of the agricultural land (Hesser et al., 2010). Since 1990 animal density has reduced by about 50% with a share of around 60% pigs and 40% cattle of total livestock units.

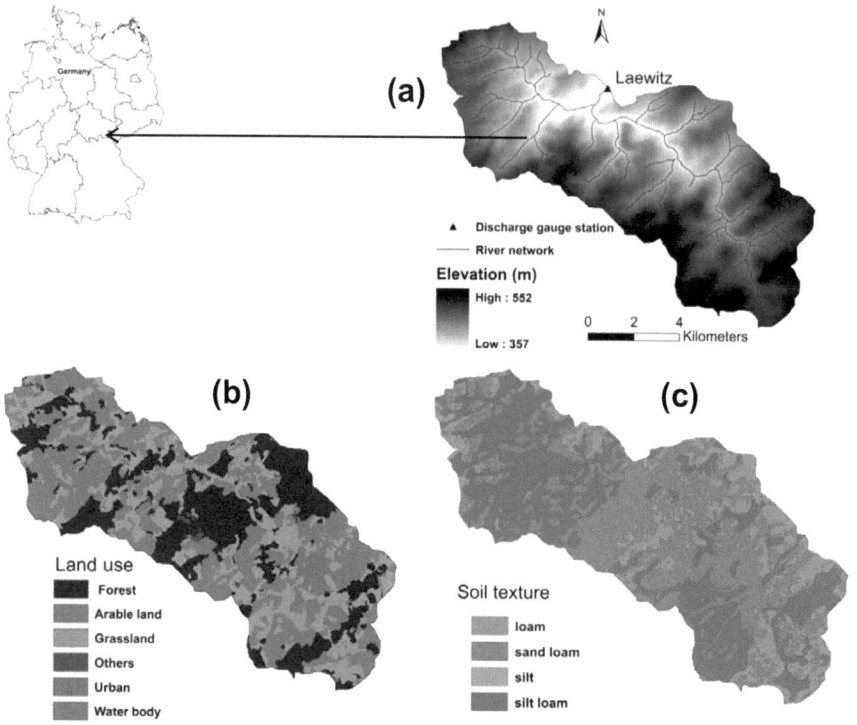

Figure 2.2 The geographical location of the Weida catchment and its Digital Elevation Model (DEM) are illustrated in panel (a). Characteristics of land use and soil are shown in panels (b) and (c), respectively.

The spatial data of the Weida catchment include a Digital Elevation Model (DEM) of 25 m grid resolution. Land use data is derived based on the classification of Landsat ETM images from the year 1999 for the entire Saale basin (Bongartz, 2004). The land use consists of six classes (Figure 2.2 b). Soil data was obtained from Fink (2004), which is based on 1:25 000 soil map and supplementary classification referring to local soil profiles. The soil consists of four main classes and 23 sub-classes (Figure 2.2 c). Weida catchment belongs to temperate climatic conditions. There are five precipitation gauging stations within/around the catchment, from which an average annual precipitation of 640 mm is estimated based on the observation between 1988 and

2004 (Balin et al., 2010; Hesser et al., 2010; Shrestha et al., 2007). Temperature, relative humidity, relative sunshine duration, global radiation, and wind speed are measured from two climate stations in the vicinity of the catchment. The average temperature is 7°C. A high resolution (1-min step) of time series river flow data from 1960 and nitrate-N data from 1997 are available at the catchment outlet Laewitz. The mean discharge observed at the Laewitz gauge station is 0.72 $m^3 s^{-1}$ (1975-2004). For nitrate measurement an online sensor with photometric principle (Stramosens CNM750/CNS70, Endress+Hausser) was used with a maximum measurement error of 2% upper range value or ± 0.1 mg l^{-1} with clear water (Hesser et al., 2010). According to the agricultural statistic of the superior administrative district Greiz, the fraction of cereals in the arable crops is 60% and the animal density in livestock units is 1.25 ha^{-1}.

Both discharge and nitrate-N concentration data show high seasonal variability. Higher discharges and nitrate-N concentrations occurred during wet period in winter between November and March, while lower discharge and nitrate-N concentrations were observed during the dry period between April and October. During the hydrological period 1998-2003, nitrate-N concentration measured at the discharge gauging station Laewitz has a range of 0.7-18.9 mg L^{-1} due to agricultural land use. The calculated mean value of nitrate-N concentration for this period based on mean daily values is 8.76 mg L^{-1}. The statistical characteristics of discharge, nitrate-N concentrations and daily nitrate-N loads for the hydrological years 1998-2003 are given in Table 2.3 (after Hesser et al., 2010).

Table 2.3 Statistical characteristics of discharge, nitrate-N concentrations and daily Nitrate-N loads for the hydrological years 1998-2003 observed at the outlet Laewitz of Weida catchment.

Variable	Unit	Period	Maximum	Minimum	Mean	Std. deviation
Discharge	$m^3 s^{-1}$	1998-2000	8.35	0.09	0.60	0.82
		2001-2003	11.4	0.01	0.60	1.04
Nitrate-N concentration	mg L^{-1}	1998-2000	18.9	0.71	8.76	3.24
		2001-2003	18.6	1.11	8.51	3.42
Nitrate-N load	kg d^{-1}	1998-2000	7580	3.96	595	969
		2001-2003	13600	5.06	642	1140

2.3 Hydrological water quality model (HYPE)

The HYPE model was developed during 2005-2007 by SMHI (Swedish Meteorological and Hydrological Institute). It was developed based on the well-known hydrologic model HBV (e.g. Bergström, 1976; Lindström et al., 1997) and hydrologic nutrient model HBV-NP (e.g. Andersson et al., 2005; Arheimer et al., 2005; Lindström et al., 2005). The schematic structure of the HYPE model is presented in Figure 2.3. In a HYPE model application, the catchment may be divided into sub-basins (Figure 2.3, left). Sub-basins can either be independent or connected by rivers and regional groundwater flow. Each sub-basin can in turn be divided into classes (Figure 2.3, left), which are the smallest computational spatial units and correspond to the hydrological response units (HRU) following Flügel (1995). The classes are not coupled to geologic locations, but defined as fractions of a sub-basin area. Land, lake and river classes are treated differently. Land classes are different combinations of soil and land use. Different vegetation types, such as forest and crops, are simulated as separate land uses. The area of a crop is constant during the simulation period, which means that crop rotation cannot be simulated.

The soil in each land class is divided vertically into one or several (up to three) layers, which may be specified with different thicknesses. The maximum water content of each soil layer is determined by three model parameters coupled to soil type and soil layer: the fraction not available for evapotranspiration, the fraction available for evapotranspiration but not for runoff and the fraction available for runoff. The sum of the three parameters comprises the maximum water content of the soil (i.e. the total porosity of the soil). The first two fractions correspond roughly to wilting point and field capacity. Each parameter can be defined as identical valuesfor different soil layersof acertain soil type. Many parameters in the model are coupled to soil or land use, while others are assumed to be general parameters in the whole catchment. Simulations start from a standard initial state, and a warming up period of typically one year is excluded from evaluation. The HYPE model simulates streamflow and nutrient concentrations in water, such as inorganic (IN) and organic (ON) nitrogen, dissolved (SP) and particulate (PP) phosphorus. In addition, it computes total nitrogen (TN) and phosphorus (TP) as the sum of the relevant fractions. Other nitrogen and phosphorus factions (e.g. humusN and humusP) are also simulated within the soil, but the runoff leaving the soil only contains the nutrient fractions mentioned above. Conservative tracers can also be modeled. Currently the model is being developed for simulation of concentration of streamwater organic carbon. A more detailed description of the hydrological and nitrogen processes in the model is given below. The equations of hydrological and nitrogen processes are given in Appendix A (after Lindström *et al.*, (2010)).

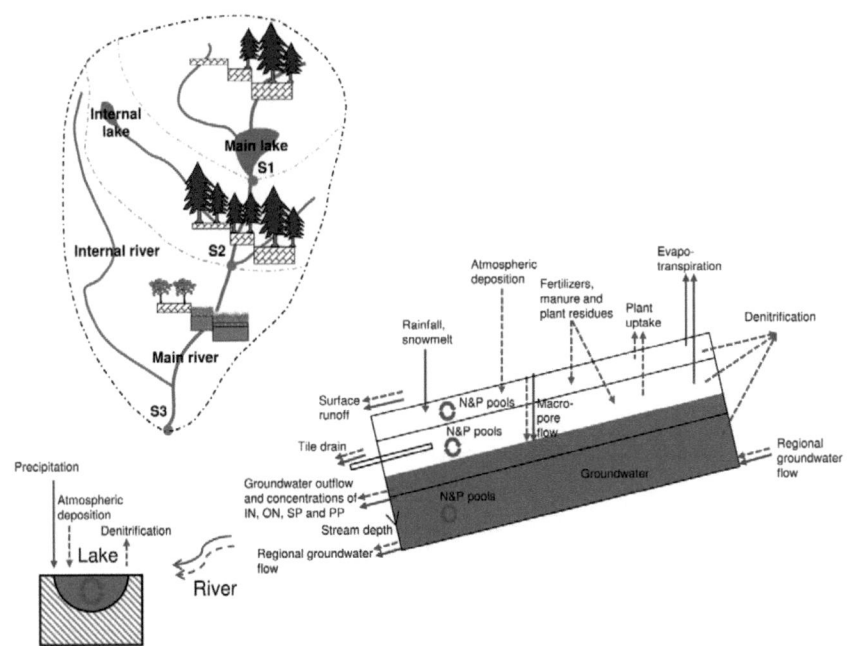

Figure 2.3 The schematic structure of the HYPE model modified from Lindström et al. (2010). The left panel shows the delineation of the catchment sub-basins system and soil land use classes (SLC), which are defined based on the DEM, land use and soil. The right panel illustrates the simulated hydrological and nutrient processes in the HYPE model. The green dash line shows the nutrient processes while the blue solid line shows the hydrological processes.

2.3.1 Hydrological processes

For hydrological simulation, the HYPE model calculates total discharge generated at the catchment outlet (m^3s^{-1}), snow accumulation and snowmelt (mm), evapotranspiration (mm), surface runoff (mm), infiltration (mm), macropore flow (mm), percolation (mm), interflow (mm), tile drain flow (mm), outflow from groundwater (losses from catchment) that corresponds to baseflow (m^3s^{-1}), river flow delay and damping (d). In addition, hydrological variables, such as soil moisture (mm), soil temperature (degrees Celsius), snow depth (cm), soil frost depth (cm), and groundwater level (m) are simulated (Lindström et al., 2010). Snow depth is estimated from the water content of snow and a snow density factor which increases with the average age of the snow pack. The soil temperature is a weighted sum of the air temperature, the soil temperature of previous day and a constant deep soil temperature. The influence of air temperature decreases with increasing snow depth. The soil frost depth is a function of soil moisture and the soil temperature. Snow depth and frost depth are estimated based on the equation developed in Lindström et al. (2002). The structure of the HYPE model for runoff process simulations are shown in Figure 2.4.

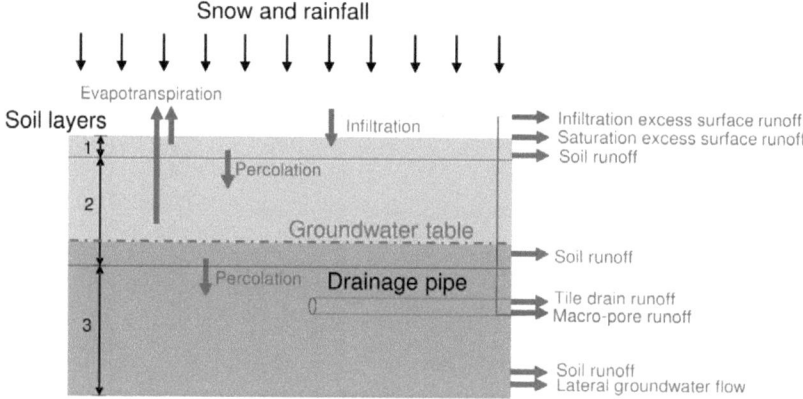

Figure 2.4 Overview of schematic structure of the HYPE model for runoff generation process simulation based on the manual of model description.

Snow accumulation and snow melt

The routine for snow accumulation and snow melt is a simplification of the HBV routine (Lindström et al., 1997 and 2010). For land classes precipitation is assumed to fall as snow below an air temperature threshold. The air temperature for each class depends on the elevation of the class. For temperatures within an interval around the threshold, a mixture of rain and snow is assumed. Snow is accumulated for each land class until melting. Snow melt is calculated with a degree-day method, and uses the same threshold temperature as snowfall. The degree-day parameter depends on land use. Storage and refreezing of meltwater and rainfall in the snow pack is not considered. The snow accumulation and snow melt are calculated following equations (1) and (2), respectively in Appendix 1A.

Surface runoff, infiltration, macropore flow and percolation

A fraction of rainfall and snowmelt infiltrates into the topsoil and the rest is diverted into surface flow and flow through macropores. Flows are diverted only if the incoming water and the soil moisture in the uppermost soil layer exceed the threshold values. Both diverted flows depend on soil type and are calculated as fractions of the incoming water above the threshold. Macropore flow primarily enters the soil layer in which the groundwater table is located. If the maximum water content of this layer is exceeded, the remaining water is stored in the layer(s) above. If the soil moisture in the uppermost soil layer exceeds the maximum water content, saturation excess overland flow occurs. It is calculated as a part of the excess water and is dependent on land use. Soil moisture above the threshold determined by the fraction of water available for runoff may percolate down to the next layer. The percolation is limited by a soil type dependent maximum percolation capacity and the available space in the soil layers below. If the maximum water content of a soil layer is reached, that soil layer is considered saturated. As shown in Figure 2.4, both infiltration excess surface runoff and saturation excess surface runoff are considered (Lindström et al., 2010). The infiltration, surface flow, macropore flow saturated overland flow and percolation are estimated following equations (3)-(7), respectively in Appendix 1.

Evapotranspiration

Evapotranspiration may occur from the two top soil layers and the depth of the second soil layer is considered as the plant rooting depth (Figure 2.4). Evapotranspiration is assumed potential if the soil moisture is above a certain threshold, and decreases linearly from that threshold to the threshold for evapotranspiration where it is zero (Lindström et al., 2010). Potential evapotranspiration depends on a land use dependent parameter (i.e. potential evapotranspiration rate), temperature, and a seasonal adjustment factor. The seasonal adjustment factor increases the evapotranspiration in spring and decreases it during autumn. It is zero for temperatures below a threshold value (same as for snowfall and snow melt). Evapotranspiration is assumed to decrease with depth. It is divided between the two layers based on evaluations of a decreasing exponential function at the midpoint of each layer. The equations for calculating evapotranspiration are shown in (8) of Appendix 1.

Interflow, tile drain flow and groundwater flow

Considering lateral subsurface water flow, interflow at different soil layers, tile drain flow and groundwater flow are simulated in the HYPE model (Figure 2.4). The soil water content in the individual soil layers determines the existence of a groundwater level. The groundwater table is located in the uppermost layer with excess water. This excess water also defines the groundwater level within a soil layer. Excess soil water can drain from any soil layer located fully or partly above the stream depth, if the soil moisture in that layer exceeds the threshold for runoff. This runoff is a fraction of the excess water determined by runoff coefficients (Lindström et al., 2010). In the soil layer where the stream depth is located, only the part of the excess soil moisture above the stream depth is included in the calculation. The runoff coefficients are coupled to the soil type, depth and slope of the soil. Tile drains are optimal in the model and the presence and depths of them are determined by the user for each land class.

The runoff from tile drains is determined by the groundwater level above the tile depth and a soil type dependent recession coefficient. The runoff is abstracted from the soil moisture in the layer in which the tile is located, and is limited to the soil moisture above the threshold for runoff in that layer. The discharge from the soil, tile drains and surface runoff are directed to the stream. Flow in the unsaturated part of the soil layers is not taken into account. Regional groundwater flow is simulated and is determined by a recession coefficient. This flow is taken from the bottom soil layer of each land class (if the soil moisture in that layer exceeds the threshold for runoff). The water may be transported to a lake at the sub-basin outlet and/or to the soil of the next sub-basin. The regional groundwater flow is added to the bottom soil layer of each land class of the receiving sub-basin. If these layers become saturated, however, the excess water is added to the layers above. Interflows, tile drain flow and regional groundwater flow are calculated following the equations (9)-(11), respectively in Appendix 1.

2.3.2 Nitrogen processes

For nitrogen simulation using HYPE, both diffuse sources and point sources are considered (Lindström *et al.*, 2010). Diffuse sources include applied organic and inorganic fertilizers, manure, plant residue and atmospheric deposition. Inorganic fertilizers and the inorganic part of manure are added to the IN pools at specified dates. The organic N fractions of manure are added to the fastN pools. The fertilizer and manure are considered to be spread in equal amounts over several days to account for uncertainties and spatial variation in application dates. The fertilizers and manure are divided and added to the top two soil layers in proportions set by the user allowing for simulation of different tillage practices. Plant residues are returned to the soil on specified days in which the nutrient contents of the residues are added to the fast pools and the slow pools. Similar as fertilizers and manure, plant residues are added to the top two soil layers. Concerning nitrogen sources from atmospheric deposition, both wet deposition and dry deposition are considered. Wet deposition is added as a concentration of IN in precipitation. Dry deposition is land-use dependent and the nitrogen is added to the IN pool in the top soil layer for land classes and to the river IN for river classes. Routines for dealing with nutrient additions from industrial, urban and rural sources are included in the model. N load from industrial and urban point sources (e.g. sewage treatment works) are added to the main river as a volume of water and concentrations of IN and ON. Nitrogen from rural households is added partly to the internal river of the sub-basin and partly to the deepest soil layers of the land classes in the sub-basin.

Nitrogen processes in both the soil profiles and the river systems are simulated in the HYPE model (Lindström *et al.*, 2010). Considering N processes in the soil, transformations between different nitrogen pools, denitrification, plant uptake and nitrogen transport through different flow pathways are simulated. For the instream nitrogen processes, denitrification, primary production and mineralization are simulated. The schematic structures of nutrient (N and P) transformation processes in the soil, river and lakes are shown in the Figure 2.5.Nitrogen transformation and sink processes in the HYPE model were described below.

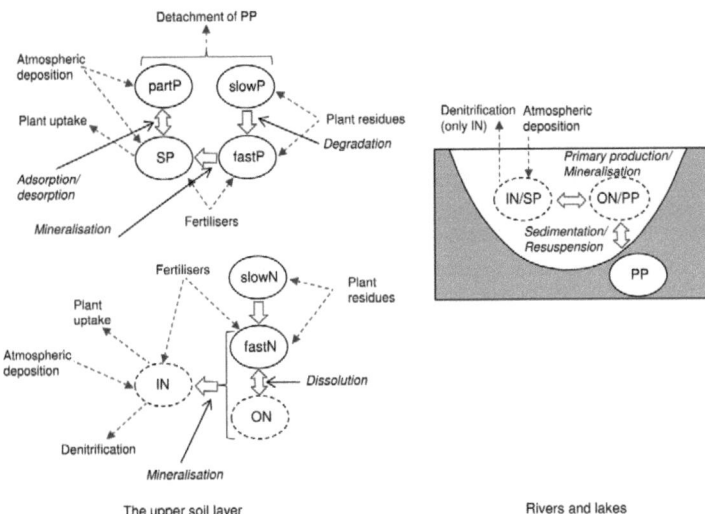

Figure 2.5 Schematic structures of nitrogen and phosphorus transformation processes in the upper soil layer, rivers and lakes and the simulated nutrient fractions (IN-inorganic nitrogen, ON-organic nitrogen, SP-dissolved phosphorus, PP-particulate phosphorus). Mobile pools are in broken lines; immobile pools in solid lines (after Lindström et al. (2010)).

Nitrogen degradation, mineralization and denitrification in the soil

In the soil, nitrogen is divided into mobile and immobile nitrogen pools. The specification of nitrogen pools and the processes affecting nitrogen in the HYPE model are similar to those described by other nutrient models, such as SOIL-N (Johnsson et al., 1987) and ANIMO (Groenendijk and Kroes, 1999).The simulated nitrogen pools consist of a slowN pool, a fastN pool, an organic N pool (ON) and an inorganic N pool (IN) (Figure 2.5). The slowN pool represents the organic nitrogen in the soil with a long turnover time. From the slowN pool, nitrogen is released to the fastN pool through degradation processes. The fastN pool represents the organic nitrogen (ON) in the soil with a short turnover time. Nitrogen in the fastN pool is available for mineralization in which it is transformed into the IN pool. Degradation and mineralization are determined by the respective transformation rate, the available amount of substrate, soil temperature, and soil moisture. The ON concentration in soil water is determined by the fastN pool. The mass transfer of nitrogen between the fastN pool and ON in solution is driven by the difference between the ON concentration in the soil solute and a steady-state equilibrium concentration calculated from the total nitrogen content of dissolved ON and the fastN pool.Denitrification is a sink of IN in all soil layers, which is defined by a function of denitrification rate, soil temperature, soil moistureand IN content of the soil. All nitrogen in the IN pool is considered to be mobile and can hence be transported between soil layers or out of the profile with horizontal and lateral soil water flow. Equations for calculating degradation, mineralization and denitrification of nitrogen in the soil are given in (12) in Appendix 1.

Plant uptake

Vegetation extracts IN and soluble phosphorus (SP) in the top two soil layers, similar to evapotranspiration (Lindström*et al.*, 2010). Potential plant uptake of nutrients in the HYPE model follows logistic uptake functions. Actual uptake only occurs between the sowing date or the start of the growing season and the harvest date or the end of the growing season. Uptake during the main growing season is limited only by the availability of nutrients. Autumn-sowing crops are limited by a temperature function during autumn to account for poorer growing conditions during this part of the year. Plant uptake of IN is calculated according to the equation (13) in Appendix 1.

Primary production, mineralization and denitrification in rivers

Primary production is a process which transforms IN into ON in water bodies. The process is reversed during mineralization. Primary production/mineralization is simulated by a function of water temperature, average total phosphorus concentration and current water temperature change. When water temperature is increasing, net primary production is assumed and when water temperature is decreasing, net mineralization is assumed. Both processes are active in the full river volume. Denitrification is a significant sink for IN in rivers. It is calculated be a function of a general denitrification rate, water temperature, IN concentration in the water body and river surface area. The primary production, mineralization and denitrification in rivers are estimated based on equations shown in (14) in Appendix 1.

2.3.3 Input data

In the model setup for runoff and nitrogen simulation using HYPE, several types of spatial data, time series data and statistical data are needed. Spatial data include Digital Elevation Model (DEM), stream network, land use and soil type. For time series data, climate forcing data of daily precipitation and daily mean air temperature for each sub-basin are required. Statistical data includes initial nutrient pools in soil, agricultural practices (i.e. manure and inorganic fertilizer applications, crop husbandry, timing and amount of fertilization, sowing and harvesting), wet and dry atmospheric depositionsof nutrient, nutrient concentrations and outflow volumes of point sources from rural households, industries and waste water treatment plants (WWTP). For model calibration and validation, discharge and streamwater nutrient concentrations (e.g. IN, ON, TN, SP, PP and TP) observed from one/several locations of the streammay be used. Observations of internal hydrological variables (e.g. groundwater level, evapotranspiration and soil moisture) can also be used for model parameter calibration and evaluation of model performance.

2.3.4 Model parameter set and parameterization

There are a relatively large number of hydrological and nitrogen process parameters required to be specified in the application of HYPE model. Model parameters are sorted into general parameters, land use dependent parameters and soil type dependent parameters. For runoff simulation, most parameters reflect water holding characteristics, evapotranspiration rates, flow paths and recession rates. Considering nitrogen simulation, the model parameters describe processes of nitrogen transformations and sinks. The hydrological water quality parameters of HYPE modelfor streamflow and streamwater IN concentration simulations are listed in Table 2.4.

Table 2.4 Descriptions and ranges of hydrological water quality parameters in the HYPE model for streamflow and streamwater IN concentration simulations.

Parameter	Description	Type	Range
Hydrological-process related parameters			
wcfc	Field capacity as a fraction, same for all soil layers (%)	Soil type dependent	0.01-1
wcwp	Wilting point as a fraction, same for all soil layers (%)	Soil type dependent	0.01-1
wcep	Effective porosity as a fraction, same for all soil layers (%)	Soil type dependent	0.01-1
cmlt	Melting parameter for snow (mm/d °C)	Land use dependent	0-10
ttmp	Threshold temperature for snow melt and evapotranspiration (°C)	Land use dependent	0-10
cevp	Evapotranspiration parameter (mm/d °C)	Land use dependent	0.01-1
rcgrw	Recession coefficient for regional groundwater flow (1/d)	General	10^{-5}-1
rrcs1	Recession coefficient for uppermost soil layer (1/d)	Soil type dependent	0.01-1
rrcs2	Recession coefficient for lowest soil layer (1/d)	Soil type dependent	10^{-5}-1
rrcs3	Recession coefficient for slope dependence (upper layer) (1/d %)	General	10^{-4}-1
srrcs	Recession coefficient for surface runoff (fraction) (-)	Land use dependent	0.1-1
trrcs	Recession coefficient for tile drains (1/d)	Soil type dependent	0.01-1
cevpam	Amplitude of sinus function that corrects potential evapotranspiration (-)	General	0.1-1
cevpph	Phase of sinus function that corrects potential evapotranspiration (days)	General	1-100
lp	Limit for potential evapotranspiration (-)	General	0.5-1
rivvel	River flow velocity (m/s)	General	0.01-10

epotdist	coefficient in the exponential function for the potential evaporations depth dependency (-)	General	1-10
Nitrogen-process related parameters			
denitr	Parameter for the denitrification in the soil (1/d)	General	10^{-5}-1
degradhn	Degradation of humus to fastN (1/d)	Land use dependent	10^{-5}-1
minerfn	Degradation n of fastN to inorganic N (1/d)	Land use dependent	10^{-5}-1
wprod	Production/degradation in water N (kg/m^3/d)	General	10^{-5}-1
denitw	Parameter for the denitrification in water (kg/m^2/d)	General	10^{-6}-1
humusN	Starting value of the soil pool of humusN (mg/m^3)	Land use dependent	0-10^6
fastN	Starting value of the soil pool of fastN (mg/m^3)	General	0-10^4
hNhalf	Half depth for humusN pool (m)	Land use dependent	0.01-1
uptsoil1	Nutrient uptake in the uppermost soil layer	Land use dependent	0.01-1
locsoil	Fraction of emission from rural waste water that is emitted directly to the lowest soil layer (rest goes to the local watercourse)	General	0-1
Qmean	Initial value for the calculation of mean flow (mm/yr)	General	0-800
fertdays	Number of days where manure is applied from the starting day the same amount every day	General	1-365

In order to reduce the risk of equifinality and over-parameterization due to model complexity and interaction between model parameters (e.g. Beven and Binley, 1992; Beven and Freer, 2001), only the sensitive parameters are chosen for the final calibration and the rest are defined as constants. The parameterization procedure follows three steps. Firstly, some parameters are defined as constants based on literature review and knowledge on physical/chemical processes (e.g. parameter for temperature's elevation dependence which is 0.6 °C per 100 m) (Strömqvist *et al.*, 2012). Secondly, the sensitive parameters are chosen and isolated from the rest less sensitive parameters through combining manual sensitivity analysis using one-factor-at-a-time approach (OFAT) and sensitivity analysis using PEST (Bahremand and De Smedt, 2010; Cuo *et al.*, 2011; Wriedt and Rode, 2006).With manual parameter sensitivity analysis (e.g. OFAT), the parameter sensitivity is evaluated by checking the effects of parameter perturbations of plus and minus a small increment (e.g. 10%) on simulated outputs (e.g. discharge or nutrient concentrations). Using PEST for parameter sensitivity analysis, the parameters' sensitivities are evaluated based on the derivatives of all observations with respect to all adjustable parameters. In the final step, the sensitive parameters obtained during the previous stepare optimized using the local search approach PEST or the global search approach DREAM$_{(ZS)}$. For parameter sensitivity analysis/calibration, the parameters' initial values and ranges (for applications using PEST) are defined according to their physical meanings, literature review, former model applications and results of premature iterative calibrations (e.g. Hesser et al., 2010; Lindström et al., 2010; Rode et al., 2009; Strömqvist et al., 2012).

2.4 Model calibration and uncertainty analysis approaches

2.4.1 PEST

PEST was developed to assist in data interpretation, model calibration and predictive analysis. PEST adjusts model parameters until the fit between model outputs and laboratory or field observations is optimized in the weighted least squares. While parameter values inferred through this calibration process are non-unique depending on parameter initial values and ranges. The repercussions of the parameter non-uniqueness on predictions can be assessed using predictive analysis mode of PEST. Moreover, PEST can be used to evaluate the effects of parameter changes on model predictive outcomes by parameter sensitivity analysis (Doherty, 2005). The theory of using PEST for parameter calibration, sensitivity analysis and predictive analysis are explained briefly below. More detailed explanation of using PEST for parameter estimation, sensitivity analysis and predictive analysis are given in Appendix 2.

2.4.1.1 Parameter estimation

PEST is a nonlinear automatic calibration tool, which offers model independent optimization routines. It uses a robust Gauss-Marquardt-Levenberg algorithm, which combines the advantages of inverse Hessian method and the steepest descent method and therefore provides faster and more efficient convergence towards the objective function minimum. PEST applies a local calibration method and searches for the best values of the model parameters by minimizing the sum of squares of the differences between measured and calculated model results (e.g. Bahremand and De Smedt, 2010; Rode et al., 2007 b). Parameter estimation using PEST is an iterative process. PEST applies a "hill-climbing" technique following the steepest gradient of the objective function starting at a specified starting point (i.e. initial value) in the parameter space (Bahremand and De Smedt, 2010; Doherty, 2005; Rode et al., 2007 b). After completing the parameter estimation process, PEST lists the optimized parameter values. It also calculates 95% confidence limits for the adjustable parameters, which are estimated on the basis of the same linearity assumption that was used to derive the equations for parameter improvement implemented in each PEST optimization iteration.

2.4.1.2 Sensitivity analysis

In the application of PEST for parameter sensitivity analysis, the value of parameter sensitivity is expressed as the relative composite sensitivity obtained by multiplying its composite sensitivity by the magnitude of the value of the parameter. It is thus a measure of the composite changes in model outputs that are incurred by a fractional change in the value of the parameter (Bahremand and De Smedt, 2010; Doherty, 2005).

2.4.1.3 Predictive analysis

When PEST is used for calibrating a model, it is asked to minimize an objective function comprised of the sum of weighted squared deviations between certain model outcomes and their corresponding field-measured counterparts, which is called running PEST in "parameter estimation mode". It is a fact that there are often many different sets of parameter values for which the objective function is at its minimum or almost at its minimum. Thus, there are many different sets of parameters, which could be considered to calibrate a model. Then it is necessary to investigate whether model generate different key model outcomes using different sets of parameter values (all of these sets being considered to calibrate the model). This question can be answered by running PEST in "predictive analysis mode". In running PEST for predictive analysis, it involves setting up a "dual model" consisting of the model run under both calibration and predictive analysis conditions. PEST's operation in predictive analysis mode has much in common with its operation in parameter estimation mode. The key model prediction generated with a parameter set calculated in this way defines the upper or lower bound of the uncertainty interval associated with that prediction.

2.4.2 DREAM$_{(ZS)}$

Global search approaches are more and more often applied for hydrologic model calibration and uncertainty analysis because of its advantages over local approaches, such as the ability to explore the whole parameter space and estimate parameters' posterior distribution functions. One of the most popular used parameter sampling approaches is the MCMC (Markov Chain Monte Carlo) algorithm. However, the problem in applying MCMC for model calibration and uncertainty analysis is its high computational cost. In the past decades, various approaches have been developed to increase the efficiency of MCMC simulation. These approaches can be categorized into single chain and multiple chain methods. Single chain methods work with a single trajectory, and continuously adapt the covariance of a Gaussian proposal distribution, using the information contained in the sample path of the chain. Two examples of self-adaptive single chain methods are the adaptive Metropolis (AM) (Haario *et al.*, 2001) and delayed rejection adaptive Metropolis (DRAM) algorithms (Haario *et al.*, 2006).

Multiple-chain methods use different trajectories running in parallel to explore the posterior target distribution. The use of multiple chains has several advantages, especially when dealing with complex posterior distributions involving long tails, correlated parameters, multimodality, and numerous optima (Laloy and Vrugt, 2012). The use of multiple chains offers a robust protection against premature convergence, and opens up the use of a wide array of statistical measures to test whether convergence to a limiting distribution has been achieved (Gelman and Rubin, 1992). DREAM (differential evolution adaptive Metropolis) algorithm is a Markov chain Monte Carlo (MCMC) approach developed by Vrugt *et al.* (2009) that uses self-adaptive randomized subspace sampling and explicit consideration of aberrant trajectories. It maintains detailed balance and ergodicity and has shown to present good performance on a wide range of model calibration studies (e.g. Dekker et al., 2012; Scharnagl et al., 2010). However, the standard DREAM requires at least N=d/2 to d (d is the dimensionality of the problem) chains to be run in parallel. Running many parallel chains is a potential source of inefficiency because each individual chain requires burn-in to travel to the posterior distribution. The lower the number of chains required, the greater the practical applicability of DREAM for computationally demanding posterior exploration problems (Laloy and Vrugt, 2012).

DREAM$_{(ZS)}$ is a globalsearch and optimization approach for Bayesian inference of the posterior probability density function of model parameters, which runs multiple different Markov chains in parallel and uses a discrete proposal distribution to evolve the sampler to the posterior distribution. It was developed based on the original DREAM algorithm but uses sampling from an archive of past states to generate candidate points in each individual chain (Laloy and Vrugt, 2012). It requires fewer function evaluations than DREAM to converge to the appropriate limiting distribution. Sampling from the past circumvents the need for a large number of parallel chains, designed to accelerate convergence for high-dimensional problems. It was proved that with DREAM$_{(ZS)}$ only three parallel chains are needed to appropriately explore the posterior pdf (probability density functions) of the calibrated parameters, reducing time required for burn-in (Schoups and Vrugt, 2010). DREAM$_{(ZS)}$ does not require outlier detection and removal, maintaining detailed balance at every single step in each of the parallel chains. In DREAM$_{(ZS)}$, the states of the chains are periodically stored in an archive using a simple thinning rule. The diminishing adaption of the transition kernel ensures convergence of the individual chains to the posterior distribution. To increase the diversity of the proposals, DREAM$_{(ZS)}$ additionally includes a snooker updater with adaptive step size. The snooker axis runs through the states of two different chains, and the orientation of this jump is different from the parallel direction update utilized in DREAM (Laloy and Vrugt, 2012).

Traditionally, estimations of parameter uncertainty and predictive uncertainty of hydrologic models assume that residual errors are independent and adequately described by a Gaussian probability distribution with a mean of zero and a constant variance. This results in the standard least squares (SLS) approaches for parameter estimation (e.g. PEST), which have an advantage that error model assumptions are stated explicitly, and their validity can be verified a posteriori (Schoups and Vrugt, 2010; Stedinger et al., 2008). However, many studies reported that, in many cases, residual errors are correlated, nonstationary (e.g. heteroscedastic), and non-Gaussian. In these cases, the residual error assumptions of formal SLS do not hold (Beven et al., 2008). Therefore, informal likelihood functions have been proposed as a pragmatic approach for uncertainty estimation in the presence of complex residual error structures, such as the likelihood uncertainty estimation methodology developed by Beven and Freer (2001). Schoups and Vrugt (2010) introduced a formal generalized likelihood function for the situations where residual errors are correlated, heteroscedastic, and non-Gaussian with varying degrees of kurtosis and skewness; through testing the new approach in runoff simulation at a humid catchment using a conceptual rainfall-runoff model, it was found that residual errors are much better described by a heteroscedastic, first-order, auto-correlated error model with a Laplacian distribution function characterized by heavier tails than a Gaussian distribution and proper representation of the statistical distribution of the residual errors yields tighter predictive uncertainty bands (Schoups and Vrugt, 2010).

2.5 Model setup and evaluation

2.5.1 HYPE model set up

2.5.1.1 Model setup at Selke catchment

The HYPE model was set up for simulations of discharge and streamwater IN concentration at the Selke catchment for a 10-year period (01.01.1994-01.01.2004). Split-sample procedure was used for model calibration (01.01.1994-01.01.1999) and validation (01.01.1999-01.01.2004). One-year simulation (01.01.1993-01.01.1994) was used for model warming up, which was excluded from model evaluation. In this study, 19 types of soil and 10 types of land use were categorized. The whole Selke catchment was divided into 29 sub-basins and 117 SLCs were defined according to the procedure described in section 2.3. The driving data of daily precipitation and daily mean temperature for discharge simulation in each sub-basin were derived from respective monitoring stations within/close to the sub-basin concerned. Agricultural practices, for instance main crop types, fertilizer inputs and sowing/harvesting dates were defined with reference to corresponding data published previously on the Selke catchment (Kistner, 2007; Kistner *et al.* 2013). The atmospheric deposition of IN (50 kg N ha^{-1} y^{-1}) was specified according to long-term measured data (Böhme *et al.*, 2003; Wriedt and Rode, 2006). Manure and plant residue amounts and their application dates were specified according to livestock in the Selke catchment and former model applications. The input data of point sources (including daily discharge and IN concentration of outflow) were defined as the respective averages of available recordings (2002-2010) from 6 sewage treatment plants within the Selke catchment.

2.5.1.2 Model setup at Weida catchment

HYPE model was set up for simulations of discharge and streamwater IN concentration at Weida catchment for a 6-year period (01.11.1997-01.11.2003).Split-sample test procedure was used for model calibration (01.11.1997-01.11.2000) and validation (01.11.2000-01.11.2003). One-year simulation (01.11.1996-01.11.1997) was used for model warming up, which was excluded from model evaluation. The whole Weida catchment was divided into 37 sub-basins and 16 SLCs were defined based on the procedures for sub-basin systems and SLCs delineation described in section 2.3. The driving data of daily precipitation and daily mean temperature for discharge simulation in each sub-basin were derived from respective five precipitation and two climate monitoring stations within/close to the sub-basin concerned. Agricultural practices, for instance main crop types, fertilizer application rates, sowing and harvesting dates were defined with reference to statistical records and previous hydrological water quality modeling studies at the Weida catchment (e.g. Hesser et al., 2010; Shrestha et al., 2007). The atmospheric deposition of IN (50 kg N ha^{-1} y^{-1}) was specified according to long-term monitored data (Böhme et al., 2003; Wriedt and Rode, 2006). Concerning the point source inputs, the data of discharge from rural households (m^3/d) and fraction of rural household N that is in soluble form are estimated based on population of the study area and reports on human water consumption etc. High temporal resolution data of time series streamflow and nitrate-N concentration measured at the catchment outlet (Laewitz) have been aggregated to daily time steps for model calibration and validation.

2.5.2 Model performance evaluation

2.5.2.1 Inorganic nitrogen loads calculation

As mentioned in the Chapter 1 (Section 1.2.6), it is important to calculate nutrient loads at different time steps for the purpose of model performance evaluation and watershed management. The measured/simulated daily IN loads were calculated by multiplying measured/simulated daily mean IN concentration with corresponding accumulated daily discharge. With this method, the variations of IN concentration and streamflow within a day are ignored. The simulated monthly IN loads were calculated as accumulation of daily loads within a month. The measured monthly loads (L) were estimated using the following interpolation method (Formula 2.1) with continuous discharge measurements and regular sampling of IN concentrations (Littlewood, 1995):

$$L = \frac{K\sum_{i=1}^{n}(C_i Q_i)}{\sum_{i=1}^{n} Q_i} \times \overline{Q_r}, \qquad (2.1)$$

where K is a conversion factor accounting for the period of load estimation and measurement units, C_i is sample concentration, Q_i is the flow at sample time, $\overline{Q_r}$ is the mean flow for the period of interest (derived from a continuous flow record).

2.5.2.2 Model performance evaluation criteria

Both graphical and statistical methods were used for model performance evaluation. Nach-Sutcliffe Efficiency (*NSE*) is widely used to assess the predictive power of hydrological water quality models.However, the *NSE* value is influenced by the sample size, outliers, bias in magnitude, time-offset bias of hydrograph models, and the sampling interval of hydrologic data (Jain and Sudheer, 2008; McCuen et al., 2006). Therefore, four statistical criteria were calculated, which are the coefficient of determination (R^2), *NSE*, Percent bias (*PBIAS*) and the *RMSE* (root mean squared error)-observations standard deviation ratio (RSR). Definitions and detailed descriptions of these criteria were given in numerous studies (e.g. Gupta *et al.*, 1999; Moriasi *et al.*, 2007; Nash and Suttcliffe, 1970; Ullrich and Volk, 2010). According to the watershed simulations evaluation guidelines given by Moriasi *et al.* (2007), the model simulation at monthly intervals can be deemed satisfactory if NSE > 0.5 and RSR < 0.70, and if PBIAS ± 25% for streamflow, and PBIAS ± 70% for nitrogen. While, model simulation at monthly intervals can be judged as good if 0.65 < NSE ⩽ 0.75 and 0.50 < RSR ⩽ 0.60, and if ± 10% ⩽ PBIAS ⩽ ± 15% for streamflow, and 25% ⩽ PBIAS ⩽ ± 40% for nitrogen. Typically, model simulations are poorer if they are evaluated at shorter time intervals (like daily) compared with longer time intervals (e.g. monthly/yearly) (Somura *et al.*, 2012). Therefore, the above guidelines can also be used for evaluation of model simulations at daily time intervals. The R^2, NSE, PBIAS and RSR were determined based on following Formulas (2.2-2.5):

$$R^2 = \left[\frac{\sum_{i=1}^{n}\left(Y_i^{obs} - \overline{Y^{obs}}\right)\left(Y_i^{sim} - \overline{Y^{sim}}\right)}{\sqrt{\sum_{i=1}^{n}\left(Y_i^{obs} - \overline{Y^{obs}}\right)^2} \sqrt{\sum_{i=1}^{n}\left(Y_i^{sim} - \overline{Y^{sim}}\right)^2}} \right] \quad (2.2)$$

$$NSE = 1 - \frac{\sum_{i=1}^{n}\left(Y_i^{sim} - Y_i^{obs}\right)}{\sum_{i=1}^{n}\left(Y_i^{obs} - \overline{Y^{obs}}\right)} \quad (2.3)$$

$$PBIAS = \frac{\sum_{i=1}^{n}\left(Y_i^{sim} - Y_i^{obs}\right) \times 100}{\sum_{i=1}^{n} Y_i^{obs}} \quad (2.4)$$

$$RSR = \frac{RMSE}{STDEV_{obs}} = \frac{\sqrt{\sum_{i=1}^{n}\left(Y_i^{sim} - Y_i^{obs}\right)^2}}{\sqrt{\sum_{i=1}^{n}\left(Y_i^{obs} - \overline{Y^{obs}}\right)^2}} \quad (2.5)$$

where, Y_i^{sim} and Y_i^{obs} are the ith simulation and observation for the variable being evaluated, respectively, $\overline{Y^{obs}}$ and $\overline{Y^{sim}}$ is the mean value of observations and simulated results for the whole period, respectively and n is the total number of observations.

2.6 Multi-site and multi-objective calibration

Multi-site and multi-objective calibration approach using PEST was employed for hydrological and water quality parameters optimization in the case study at Selke catchment. Here, multi-site calibration means spatially distributed discharge/IN concentration observations are used in parameter optimization; multi-objective calibration means calibrating hydrological and water quality parameters simultaneously using all relevant observations. The global objective criterion was defined as weighted sum of different objective functions by referring to van Griensven and Bauwens (2003). Each objective function corresponds to sum of squared deviations between simulated and measured variables (discharge/IN concentration) at each gauging station. In order to ensure that all the objective functions are of the same order of magnitude and therefore have similar significance in the search for the optimum, objective functions were weighted. The weight of each observation group was assigned as reciprocal of its standard deviation. The definition of global objective criterion and objective function are given in Formulas (2.6) and (2.7), respectively:

$$GOC = \sum_{i=1,m} \omega_i OF_i \quad (2.6)$$

and

$$OF = \sum_{j=1,n} [x_{j,measured} - x_{j,simulated}]^2 \quad (2.7)$$

where, m corresponds to the total number of observation groups of measured discharge and IN concentrations from the three gauging stations, n corresponds to the total number of measured discharge/IN concentrations at each gauging station and ω is the weight of the concerned objective function.

2.7 Step-wise calibration and multi-objective calibration

As mentioned in Chapter 1 (Section 1.2.5), in integrated hydrological water quality modelling step-wise calibration means calibrating the hydrological parameters first and then calibrating the water quality parameters while keeping the optimized hydrological parameters from first step as constants. In multi-objective calibration, hydrological and water quality parameters are calibrated simultaneously using both corresponding observations (e.g. streamflow and nutrient concentration measurements). In this study both step-wise (SWC) and multi-objective (MOC) calibrationwere implemented for hydrological water quality parameter optimization of the HYPE model at Selke and Weida catchments. Modelling results derived from SWC and MOC in terms of parameter posterior uncertainty ranges, model performance on discharge and IN simulations, and robustness of optimized parameter set were compared. To this end, PEST was combined with HYPE model for parameter automatic calibration.

In the Weida catchment case study continuous daily mean discharge and nitrate-N concentration measurementsfrom catchment outlet Laewitz for the period 01.11.1997-01.11.2000 were used for model calibration; the model was evaluated during the period 01.11.2000-31.12.2003. In the Selke catchment case study observations of continuous daily mean discharge and non-continuous regular IN concentrations from all three gauging stations (Silberhuette, Meisdorf and Hausneindorf) for the period 01.01.1994-01.01.1999 were used for model calibration to benefit fromthe spatially distributed observations; the model was validated during the period 01.01.1999-01.01.2004.Selke and Weida catchments have various physiographical characteristics, which results in different dominant hydrological, nitrogen transport and turnover processes. Therefore, the parameters chosen for calibration in these two catchments were not the same according to parameter sensitivity analysis. However, under each case study parameters and their initial values and ranges defined for SWC and MOC are identical in order to make objective comparison.

2.8 Effects of spatial and temporal resolution of calibration data on model identification

In order to evaluate the effects and information contents of different spatial and temporal resolution of calibration data (streamflow and IN concentration observations) on model identification, both PEST and DREAM$_{(ZS)}$ were coupled with the HYPE model to implement calibration and predictive analysis at Selke catchment. In this part we presented the schemes for investigating the effects of spatial resolution of discharge and IN concentration observations, temporal resolution of IN concentration observations, as well as comparing model identification results derived from PEST and DREAM$_{(ZS)}$.

2.8.1 Single-site vs. Multi-site calibration

To investigate the effects of spatial resolution of calibration data (i.e. discharge and IN concentration observations) on model identification, single-site calibration and predictive analysis (SSCP) that uses the observed discharges/IN concentrations only from catchment outlet and multi-site calibration and predictive analysis (MSCP) that uses the observed discharges/IN concentrations from both catchment outlet and internal sites were implemented and compared. Hydrological and nitrogen-process parameter calibration and uncertainty analysis were implemented on the period 01.01.1994-01.01.1999. The performance of optimized parameter sets were evaluated for the period 01.01.1999-01.01.2004. The schemes for comparing SSCP and MSCP were illustrated in Figure 2.6.

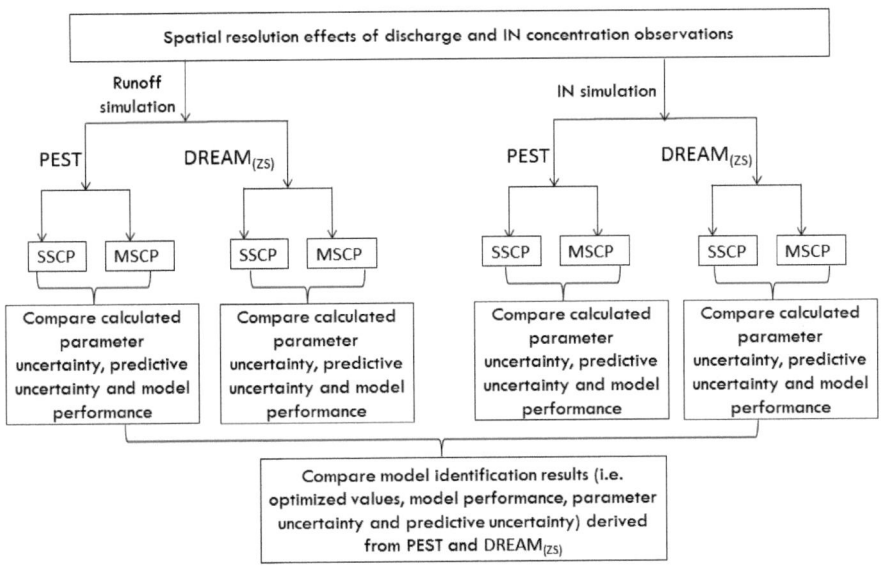

Figure 2.6 Schemes for investigating the effects of spatial resolution of the calibration data (discharge and IN concentration observations) on respective hydrological and nitrogen-process parameter identification. SSCP and MSCP stand for single-site calibration and predictive analysis and multi-site calibration and predictive analysis, respectively.

2.8.2 Bi-weekly vs. Daily IN measurements

To investigate the effects of temporal resolution of calibration data (IN concentration observations) on model identification, nitrogen-process parameter calibration and uncertainty analysis using respective biweekly nitrate-N concentration measurements (BWSC) and daily averages of nitrate-N concentration observations (DAOC)of the period 01.11.2010-31.12.2011were implemented and compared. The model performance using the optimized parameter sets were validated for the period 01.01.2012-31.12.2012.The procedures to compare model identification results against BWSC and DAOC are shown in Figure 2.7.

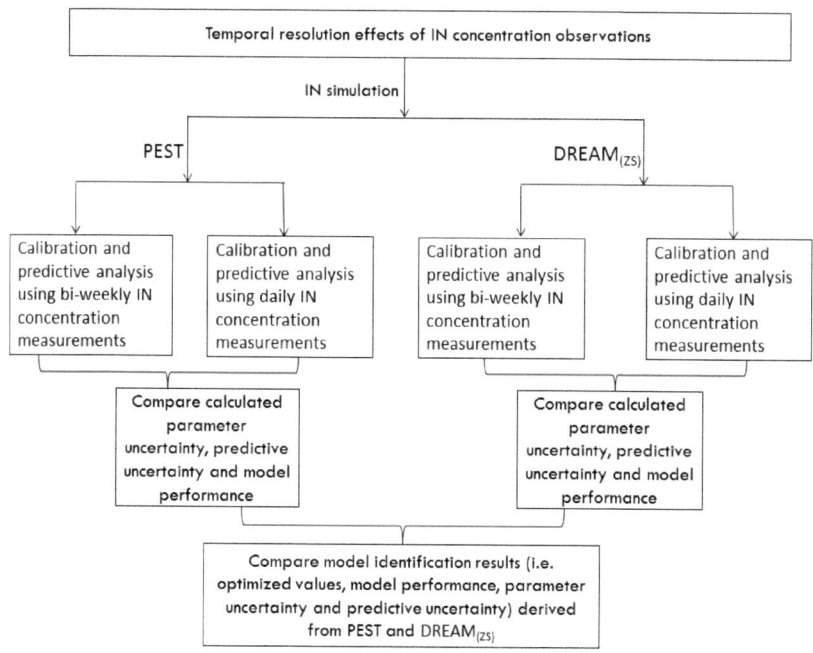

Figure 2.7 Schemes for investigating the effects of temporal resolution of the calibration data (IN concentrations) on water quality parameter identification.

Chapter 3: Multi-site and multi-objective calibration of integrated catchment model HYPE

In this chapter, HYPE model was applied in Selke catchment to simulate streamflow, IN concentration and IN load. PEST was combined with the HYPE model to implement multi-site and multi-objective calibration of hydrological water quality parameters. The objectives of this study are (i) to assess the capability of HYPE model for simulating runoff and IN leaching in nested and spatially heterogeneous mesoscale catchments in central Germany, and (ii) to investigate the temporal and spatial variations of IN leaching related to catchment heterogeneity (e.g. precipitation, topography, land use and soil type).

3.1 Model calibration, hydrological and IN simulation results at Selke catchment

3.1.1 Model parameter calibration results

The parameter calibration results using PEST and their physical meanings are listed in Table 3.1. Based on the values of parameters' relative composite sensitivity, it was found that the most sensitive hydrological parameters are *wcep* (for dominant soil type in the mountain areas-cambisols), *rivvel* (general paramter), *cevp* (for dominant land use-arable land and forest) and *epotdist* (general paramter), in decreasing order. The most sensitive parameter is *wcep*, which defines the fraction of soil volume available for soil runoff. The high sensitivity of parameter *wcep* for cambisols (relative composite sensitivity of 0.0062) indicates the importance of subsurface runoff in the mountain areas. The parameter *rivvel* describes the river flow velocity, which is sensitive for regional hydrograph representation. The parameters*cevp* for the two dominant land use classes (arable land and forest) and *epotdist* are sensitive as they determine the evapotranspiration. For the nitrogen processes, parameter uptsoil1 (for the dominant land use-arable land) is most sensitive, which determines the fraction of plant uptake from the first soil layer (relative composite sensitivity is 0.011). The parameter *denitr* is also sensitive as it defines the denitrification rate of IN in the soil, which is another important sink of nitrogen. Results showed that the parameters related to nitrogen processes in the soil (*denitr* and *uptsoil1*) are more sensitive than those parameters relevant to in-stream processes (*denitw*, *wprod* and *rivvel2*). Parameter 95% confidence limits become narrower with increasing parameter sensitivity (Table 3.1). Results showed that the final optimized values are close to the initial values. This can be explained by the local search nature of PEST and the proper choice of initial values and ranges according to the calibration procedure described in Chapter 2 (Section 2.4.1.1).

Table 3.1 Physical meanings, initial values and ranges, parameters' sensitivity, optimized values and confidence limits of key parameters. Shaded rows are the parameters related to hydrological processes, while the remaining are parameters related to nitrogen processes.

Parameter	Physical meaning	Initial value	Initial range	Relative composite sensitivity	Optimized value	95% confidence limits
cevp						
Agriculture land	Potential evapotranspiration rate (mm d^{-1} °C^{-1})	0.22	0.01-1	0.0054	0.234	0.219-0.249
Coniferous forest		0.16	0.01-1	0.0045	0.170	0.159-0.182
Mixed forest		0.12	0.01-1	0.0049	0.116	0.110-0.121
rrcs1						
Cambisols	Soil runoff coefficient for the uppermost soil layer (d^{-1})	0.105	0.001-1	0.0016	0.104	0.089-0.119
rivvel	Maximum velocity in the stream channel (m s^{-1})	0.20	0.001-1	0.0058	0.202	0.196-0.207
rcgrw	Runoff coefficient for regional groundwater flow (d^{-1})	0.0043	0.0001-0.1	0.0023	0.0045	0.0038-0.0051
epodist	Decrease of evapotranspiration with soil depth (m^{-1})	6.646	1-10	0.0041	6.574	6.04-7.10
wcfc						
Cambisols	Fraction of soil layer where water is available for evapotranspiration but not for runoff (-)	0.7	0.01-0.7	0.0021	0.7	0.599-0.80
wcep						
Cambisols	Fraction of soil layer where water is available for runoff (-)	0.133	0.001-0.2	0.0062	0.12	0.11-0.13
pcadd	Correction parameter for precipitation (-)	0.087	0.001-1	0.0021	0.0942	0.0781-0.110
denir	Denitrification rate in soil (d^{-1})	0.0243	0.001-0.1	0.0026	0.0228	0.0181-0.0275
uptsoil1						
Agriculture land	Fraction of nutrient uptake in the uppermost soil layer (-)	1.0	0.001-1.0	0.011	1.0	0.96-1.04
Coniferous forest		1.0	0.001-1.0	0.002	1.0	0.80-1.20
Mixed forest		1.0	0.001-1.0	0.0028	0.94	0.83-1.05
deniw	Parameter for the denitrification in water (kg m^{-2} d^{-1})	1×10^{-5}	1×10^{-6}–0.1	6.9×10^{-7}	1×10^{-6}	-5.35×10^{-4}–5.37×10^{-4}
wprod	Production/decay of N in water (kg m^{-3} d^{-1})	0.0088	0.0001-0.1	0.0004	0.0551	0.0042-0.011
rivvel2	Parameter for calculating the velocity of water in the stream channel (-)	0.928	0.001-1.0	0.0004	1.0	0.629-1.371

3.1.2 Hydrological simulation

The discharge simulation results and the model performances at the stations Silberhuette, Meisdorf and Hausneindorf are shown in Figure 3.1a and Table 3.2, respectively. The HYPE model reproduced the temporal variations of discharge good during both calibration and validation periods at all three stations, with lowest NSE of 0.86 (Table 3.2). At station Silberhuette, the model underestimated the measured discharge during the validation period (PBIAS = - 10.3%), which is greater than the under-prediction for the calibration period (PBIAS = - 4.9%). This is probably attributed to inter-annual climate variation (precipitation and temperature) and input data error (i.e. underestimation of precipitation). At the station Hausneindorf, the effects of streamflow extraction on discharge during the period Nov. 1998- Dec. 2004 were taken into account by adding the mean pumping rate ($\approx$ 0.18 m^3 s^{-1}) and average groundwater flow ($\approx$ 0.09 m^3 s^{-1}), which were estimated based on monthly pumping amount and change of lake volume to the original recorded discharge. The average pumping rate reflected around 18% of the mean streamflow during the validation period (1999-2004). The ability of the HYPE model to capture the effect of streamwater withdrawal on the discharge simulation was tested through comparing the PBIAS calculated by ignoring/considering the streamwater extraction. The difference was evaluated as 15%. This is approximately equal to the streamwater withdrawal amount (15% vs. 18%) considering that the pumping volume varied over time, which affirms the ability of the HYPE model to capture this artificial impact. Some peak flows in winter were underestimated, probably due to the simplified description of the snowmelt process in the HYPE model. Moreover, the model calculates streamflow at daily intervals so that it cannot predict precisely the flood events, which are caused by high intensity and short-term rainfall events. The underestimation of peak flows was also reported in former studies (e.g. Lam *et al.*, 2012; Rode *et al.*, 2009).

The ability of the HYPE model to predict the internal hydrological variable (soil water storage) was investigated. The HYPE model was able to reproduce the hydrograph during two extreme climatological events (wet year of 2002 and dry year of 2003 with yearly precipitation of 786 and 395 mm, respectively). The NSE are 0.89 for the year 2002 and 0.92 for the year 2003. In particular, the model represented the peak flow observed in August 2002 and the low flow obtained during the same month of the drought year 2003 (Figure 3.1b). The difference in mean soil water storage between August of 2002 and August of 2003 was about 73 mm. Soil water storage anomalies (Seneviratne *et al.*, 2012), which is computed as the difference in percentage between the predicted monthly average soil water storage (integrated value for the whole soil column) during the period 2002-2004 and the corresponding monthly average values during the whole simulation period of 1994-2004, is shown in Figure 3.1b. The trend of soil water storage anomalies during the period of 2002-2004 are consistent with that derived from lysimeter measurements in Swiss pre-alpine areas (Seneviratne *et al.*, 2012). The soil water storage anomalies in August of 2002 and August of 2003 are 3.0% and - 5.3%, reflecting anomaly wet and dry conditions, respectively. The difference in order of magnitude between results obtained in this study and findings presented by Seneviratne et al. (2012) can be explained by the difference in precipitation between these two studied catchments. The HYPE model simulated the discharge good at both calibration and validation periods for all three stations as the lowest performances obtained at Hausneindorf (NSE = 0.86, RSR = 0.37 and PBIAS = 14.3%). The model performance decreased slightly from headwater to catchment outlet (Table 4). The underestimation in terms of water balance in headwater and the over-prediction in lowland area is probably attributed to simplified evapotranspiration routine in the HYPE model, which cannot sufficiently represent the processes driving the spatial variation of evapotranspiration (Strömqvist et al., 2012).

Table 3.2 Model evaluation statistics for discharge, daily and monthly IN load simulations at the gauging stations Silberhuette, Meisdorf and Hausneindorf during the calibration (1994-1999) and validation (1999-2004) periods (PBIAS has the unit %, the other criteria are unitless).

Variable	Criterion	Calibration			Validation		
		Silberhuette	Meisdorf	Hausneindorf	Silberhuette	Meisdorf	Hausneindorf
Discharge	R^2	0.89	0.88	0.86	0.92	0.90	0.88
	NSE	0.88	0.88	0.86	0.91	0.90	0.86
	PBIAS	- 4.9	- 3.8	2.6	- 10.3	- 0.7	14.3
	RSR	0.34	0.35	0.37	0.31	0.32	0.37
Daily IN load	R^2	0.91	0.82	0.72	0.83	0.91	0.83
	NSE	0.88	0.80	0.70	0.83	0.89	0.46
	PBIAS	- 11.4	- 15.3	4.3	0.5	8.1	40.3
	RSR	0.34	0.45	0.54	0.41	0.34	0.74
Monthly IN load	R^2	0.91	0.85	0.83	0.86	0.90	0.87
	NSE	0.85	0.76	0.78	0.86	0.90	0.69
	PBIAS	- 15.6	- 21.6	- 6.7	0.2	- 1.0	33.2
	RSR	0.39	0.49	0.47	0.38	0.31	0.56

The area-weighted annual runoff measured during the 10-year simulation period (1994-2004) are 415 mm, 265 mm and 133 mm for the sub-basin Silberhuette, sub-basin Meisdorf and the whole Selke catchment, respectively. This indicates that runoff is mainly produced in the headwater (sub-basin Silberhuette). In response to the spatial variability of topography and soil type the simulated flow components exhibited significant decrease in the contribution of interflow, in the upper stream mountain areas the interflow contribution was at 80% whereas in the downstream area it was at 16%. The high share of interflow in headwater is consistent with the finding derived from former runoff generation study conducted in Schaefertal catchment which has similar topography, geological condition and soil type (Becker and Mcdonnell, 1998). The change of hydrological regimes related to the spatial variability of the catchment characteristics is also consistent with findings from former study (Chen et al., 2010), in which it was reported that topography plays a dominant role in formation of runoff components. For instance, increase of catchment slope could increase the subsurface storm flow whilst decrease the overland flow and base flow. Also, a decrease of the lower layer hydraulic transmissivity may result in increase of the overland flow and sub-surface storm flow and decrease of base flow.

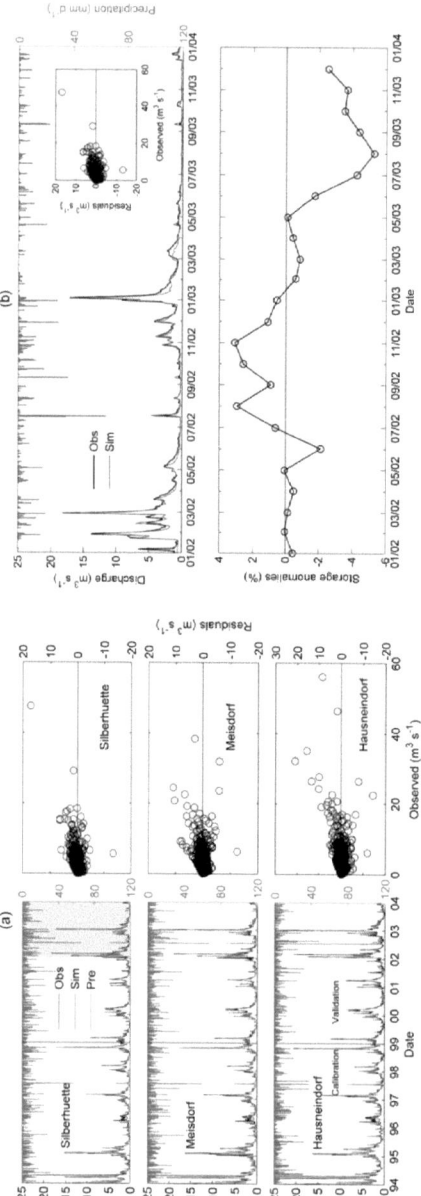

Figure 3.1 Simulated (Sim) and observed (Obs) streamflow at the gauging stations Silberhuette, Meisdorf and Hausneindorf as well as precipitation (Pre) for calibration (1994-1999) and validation (1999-2004) periods together with plots of residuals are shown in panel (a). The model's performance during two extreme climatological events occurred in the period 2002-2004 (gray area in panel (a)) at the station Silberhuette is illustrated in panel (b). The latter shows, moreover, the soil water storage anomalies (%) from January of 2002 to December of 2003.

3.1.3 IN simulation

3.1.3.1 IN concentration and daily IN load

The observed and predicted IN concentration and daily IN load for all three stations are illustrated in Figures 3.2 and 3.3, respectively. The model performances of daily IN load simulations are given in Table 3.2. Observed IN concentrations show similar pattern at the stations Silberhuette and Meisdorf in terms of timing and magnitude. This can be explained by the similar land use (forest dominant) and hydrological regimes (interflow dominant) between these two sub-basins. In addition, observed IN concentrations reveal consistent seasonal variation during the entire simulation period, which is represented by high and low concentrations in the winter and summer, respectively. The seasonal pattern of IN concentrations is the results of combined effects of hydrological and biogeochemical processes on IN transport and retention. In winter, higher share of interflow transports more nitrogen into the stream with short residence time. In terms of biogeochemical processes, denitrification and plant uptake are low in winter due to the low temperature. On contrary, the lower IN concentrations during low-flow condition in summer are explained by higher share of base flow and high retention (e.g. denitrification in soil and groundwater) as well as increased plant uptake due to high temperature.

The calculated plant uptake and denitrification in the soil showed seasonal dynamics, which is characterised by high values in summer and low values in winter. This verifies that the HYPE model is able to capture the nitrogen transport and transformation processes and dynamics related to variable climate and flow conditions. This is also indicative of the significant impact that hydrological processes have on streamwater IN concentration. For the catchment outlet (Hausneindorf), however, measured IN concentrations are characterized by smaller fluctuation compared with the two upstream stations (Figure 3.2). This is probably attributed to (i) the increased catchment size which dampens the dynamics of IN concentration and (ii) the possible inputs of point sources in the lower part of the catchment. Concerning the temporal variation, the daily IN load showed similar seasonal dynamics as IN concentration, which is the combined result of seasonality of discharge and IN concentration (Figure 3.3). The daily IN loads at station Hausneindorf are higher – compared with the upper-stream stations – due to the elevated fertilizer input from arable land, which represents about 76% of the lowland areas.

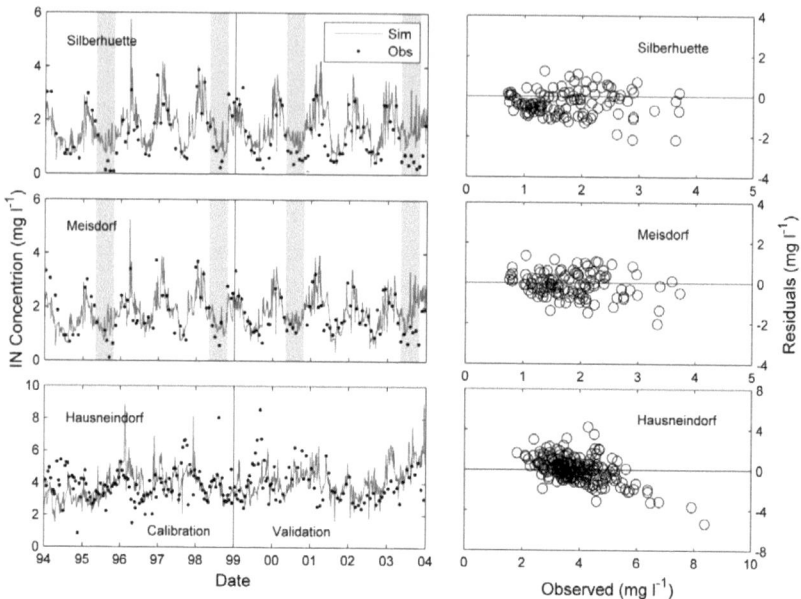

Figure 3.2 Simulated (Sim) and observed (Obs) IN concentrations together with plots of residuals at the stations Silberhuette, Meisdorf and Hausneindorf for the calibration (1994-1999) and validation (1999-2004) periods. Marked grey areas show overestimation of IN concentration during summer low-flow conditions.

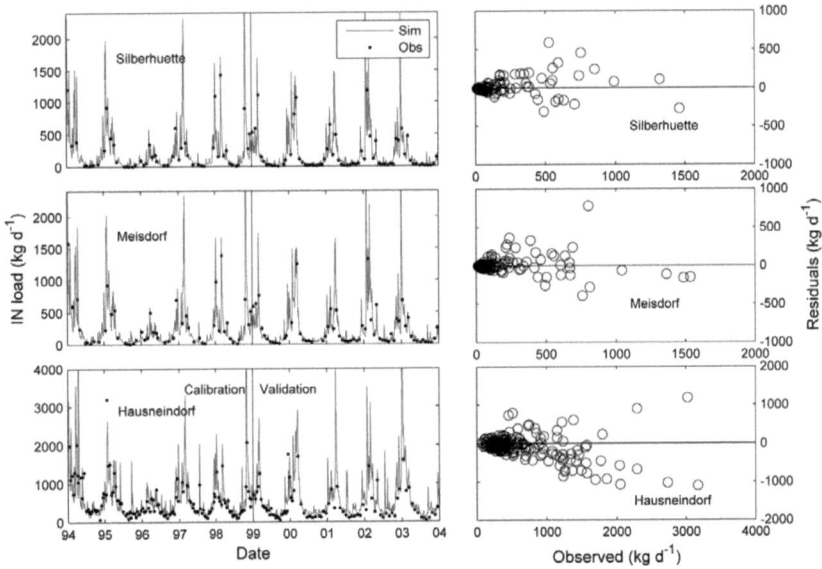

Figure 3.3 Simulated (Sim) and observed (Obs) daily IN load together with plots of residuals at the stations Silberhuette, Meisdorf and Hausneindorf during the calibration (1994-1999) and validation (1999-2004) modes.

The simulated IN concentrations represent the dynamics of the measured values. However, the model overestimated the IN concentrations during low-flow conditions in summer (Figure 3.2). The measured daily IN loads are reproduced very good during both calibration and validation periods for the stations Silberhuette and Meisdorf (lowest performance at the station Meisdorf, NSE = 0.80 and PBIAS = -15.3%). For the station Hausneindorf, the model represented the measured daily IN loads good only during the calibration period (NSE = 0.70 and PBIAS = 4.3%). During the winter periods, some daily IN load peaks are under-predicted (e.g. the high-flow events occurred at the end of January 1995 and February of 1997 at the station Hausneindorf), which is probably due to underestimation of peak flow. Concerning low-flow periods, the model generally overestimates the measured daily IN loads due to over-prediction of IN concentration as mentioned above. The overestimation of IN daily load is more pronounced during the validation phase at the station Hausneindorf due to discharge measurement error caused by the artificial streamwater extraction (Figure 3.3). The discrepancies between simulated daily IN loads and corresponding measurements are mainly attributed to the mismatches between observed and predicted discharge, which indicates the importance of good hydrological simulation for representation of IN loads.

3.1.3.2 Monthly and yearly IN loads simulations

Figure 3.4 shows simulated and measured monthly IN load for both calibration and validation periods at the stations Silberhuette, Meisdorf and Hausneindorf. The seasonal dynamics of the monthly IN load are captured by the model consistent with the daily IN load simulations. The model's performances are good for all three stations during both calibration and validation periods as NSE > 0.65, RSR < 0.60 and PBIAS < ± 40% (Table 3.2). The measured and predicted mean monthly IN load for all three stations are presented in Figure 3.5. Results showed good model performances at the two upper-stream stations during both calibration and validation modes. For the lowland station (Hausneindorf), the model represents good the measured average monthly IN load only during calibration phase. For the validation mode, however, the monthly IN load is consistently overestimated due to streamwater extraction (Figure 3.5). Following the similar procedure of estimating the streamwater extraction effect on discharge described in section 3.1.2, the streamwater extraction may result in underestimation of monthly IN load by 1.758 t month^{-1}, decrease of NSE by 0.15 and increase of PBIAS by 18.3% if the underestimation of observed discharge is not compensated.

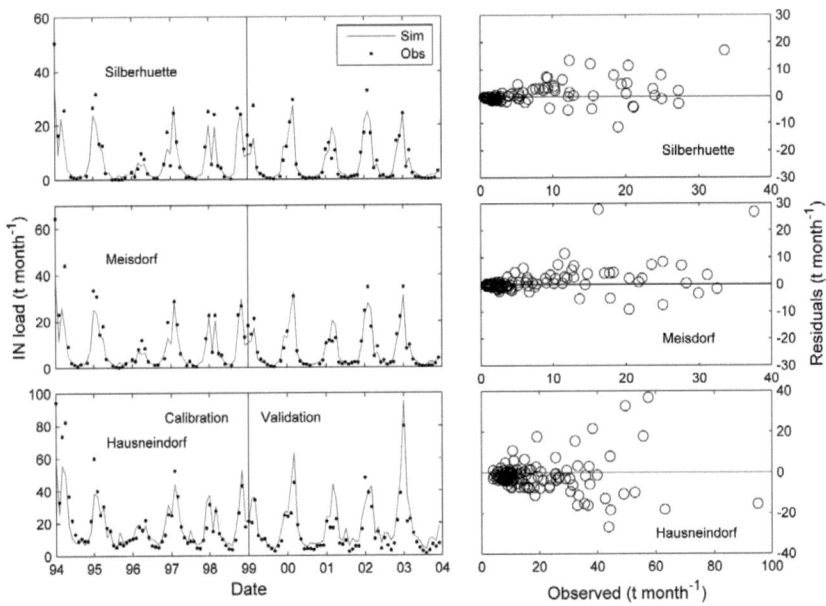

Figure 3.4 Simulated (Sim) and observed (Obs) monthly IN load together with plots of residuals at the stations Silberhuette, Meisdorf and Hausneindorf for the calibration (1994-1999) and validation (1999-2004) periods.

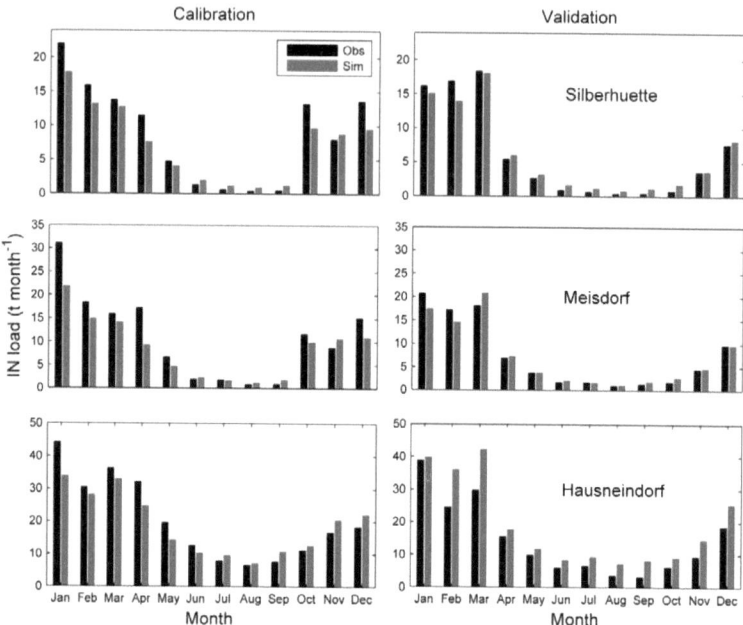

Figure 3.5 Simulated (Sim) and observed (Obs) average monthly IN load at the stations Silberhuette, Meisdorf, and Hausneindorf for the calibration (1994-1999) and validation (1999-2004) periods.

To investigate the spatial variation of IN leaching within the Selke catchment, the average area-weighted annual IN leaching load from each sub-basin, during calibration and validation modes, were calculated and presented in Figure 3.6. The IN leaching load has a range of 0.01-9.23 kg ha^{-1} y^{-1} and 0.06-10.55 kg ha^{-1} y^{-1} for the calibration and validation periods, respectively. Sub-basin 14, which is located in the lowland areas, has the highest IN leaching load, which is 9.23 kg ha^{-1} y^{-1} for the calibration mode and 10.55 kg ha^{-1} y^{-1} for the validation mode. As mentioned above, this can be attributed to the large share of arable land (71% of the sub-basin area) and consequently high nutrient inputs from fertilizer application. The sub-basins 21, 24, 25, 26 and 28, which are located in the upper-stream mountain areas, show also high IN leaching loads, having ranges of 7.40-8.73 kg ha^{-1} y^{-1} and 6.74-7.60 kg ha^{-1} y^{-1} for the calibration and validation periods, respectively (Figure 3.6). This can be explained by high nutrient transport capacity through interflow corresponding to steep slope and highly permeable soils (cambisols) in the upland area. In these sub-basins, forest is the dominant land use, whose fraction varies between 54-82% relative to arable land covering 10-42% of the sub-basin area. Among these five sub-basins, sub-basin 21 has the lowest annual IN leaching load due to its relatively small share of arable land (9.5% of the sub-basin area). Sub-basin 26 shows high values (quantified by 8.16 kg ha^{-1} y^{-1} and 7.34 kg ha^{-1} y^{-1} for the calibration and validation periods, respectively), although 82% of this sub-basin is covered by forest. This is attributed to its steep slope (mean slope of the sub-basin being 5°) and large share of highly permeable soils-cambisols (63%), which results in a large nutrient transport capacity through subsurface flow.

Sub-basins 1, 12 and 18 located in the lowland areas are dominated by arable land (being 75-100% of the corresponding sub-basin area). However, they represent low IN leaching loads, having ranges of 0.01-1.54 kg ha^{-1} y^{-1} and 0.06-1.60 kg ha^{-1} y^{-1} for the calibration and validation periods, respectively. This is explained by the low nutrient transport capacity due to small share of interflow caused by low precipitation, high evapotranspiration, mild slope and dominated low permeable soils-chernozems, although nutrient inputs from fertilizer application are high. Sub-basin 20 is nearly totally covered by forest (99%), which has low nutrient inputs resulting in low IN leaching load. Based on the spatial variation of the annual IN leaching load, it is concluded that the IN leaching load is dependent on both runoff and land use.

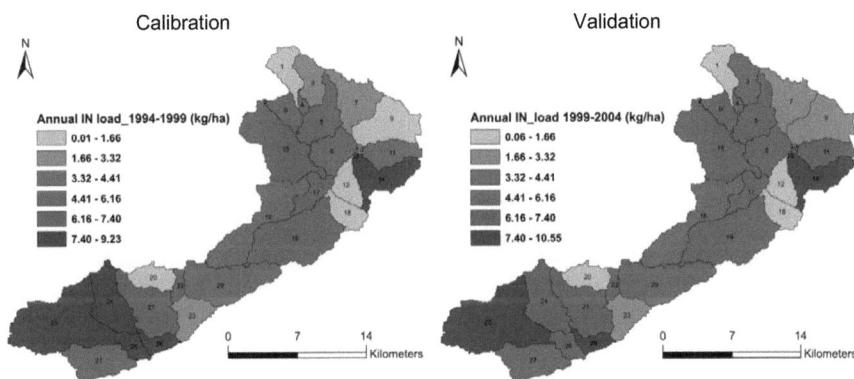

Figure 3.6 Simulated average area-weighted annual IN load delivered to stream by sub-basins of the Selke catchment during the calibration (1994-1999) and validation (1999-2004) periods.

3.2 Discussion

Compared with the classical step-wise calibration approach, calibrating hydrological and water quality parameters simultaneously increases constraints on both processes. The multi-site calibration is challenging at a heterogeneous nested catchment due to the interaction between sub-basins. It was reported that model structure of distributed hydrological models were only transferable within catchments which have similar initial soil moisture conditions and catchment topography (e.g. Hunukumbura et al., 2012). Using multi-site and multi-objective calibration approach, the HYPE model reproduced good the hydrographs at all three discharge gauging stations (Silberhuette, Meisdorf and Hausneindorf) (Table 3.2). Additionally, the model reproduced well hydrographs and soil water storage anomalies during the extremely wet year of 2002 and the extremely dry year of 2003 occurred in central Europe. This indicates that the HYPE model is able to represent the spatial variation of hydrological regimes in heterogeneous catchments.

The different dynamics of IN concentration across the three gauging stations highlight the combined effects of land use and hydrological regimes on nitrogen transport. The decrease in seasonal variation of IN concentration from the headwater to lower parts is mainly attributed to different IN retention patterns (Figure 3.2). A positive relationship between nutrient in-stream retention and sub-basin slope was reported in former studies (Grizzetti et al., 2003; Wondzell, 2011). The seasonal variation of IN concentration can be explained by the different nutrient transport capacities and the magnitude of IN retention as well as plant uptake variability throughout the year. Ye et al. (2012) found that dissolved nutrient retention on the channel network scale is influenced by flow conditions, streamflow variability and catchment scale. Creed et al. (1996) suggested flushing mechanism and draining mechanism during low soil saturation deficit and high soil saturation deficit periods, respectively. Shrestha et al. (2007) found a negative relationship between temperature and streamwater nitrate-N concentration attributed to the temperature effects on runoff and nitrate-N in-stream retention. In-stream denitrification tends to decrease from headwater to lower part due to decreasing hyporheic flow exchange (Wondzell, 2011). In the HYPE model, the nitrogen retention processes (e.g., denitrification in the soil) are highly dependent on the user-defined general reaction coefficients. However, the effects of initial soil moisture content, flow condition and catchment scale are not fully considered in the model. In the HYPE model in-stream denitrification is computed considering only denitrification rate, the size of wetted perimeter of the river cross-section, IN concentration and the streamwater temperature. The in-stream denitrification rate is considered as a general and constant parameter. However, the effects of stream reach characteristics, such as flow velocity or sediment properties, on denitrification are not considered in the model. The nitrogen in-stream retention processes in HYPE model need to be improved.

Our finding showed that 95% posterior uncertainty ranges of the sensitive hydrological parameters decreased when multi-objective calibration was employed compared with step-wise calibration (unpublished data). This indicates that multi-objective calibration improves hydrological parameter identification. Moreover, to investigate the effect of parameter non-uniqueness on prediction, predictive analysis was implemented on both discharge and IN concentration simulations using PEST (Doherty, 2005). It was found that parameter uncertainty does not result in high level of predictive uncertainty, especially very low predictive uncertainty ranges were found for discharge simulation (unpublished data). However, it needs to be mentioned that PEST is a local search approach. Thus, the results of parameter calibration (i.e. the stability of the final optimized values) and predictive uncertainty analysis depend greatly on parameter initial values and ranges defined by the user. In future, a global approach (i.e., Monte Carlo Markov Chain) will be used for parameter and prediction uncertainty analysis (Laloy and Vrugt, 2012).

With regard to catchment management, the overestimation of the IN load during low-flow conditions in summer is less significant compared with good predictions during high-flow conditions in winter because nitrogen is mainly transported during high-flow. Considering spatial variation of the area-weighted IN leaching load, overall higher values in the headwater mountain areas are noted compared with the lowland areas, although higher IN concentration was observed in the downstream (Figure 3.6). This is explained by the substantial decrease of interflow due to spatial variability of climate patterns, topography, land use and soil type. Therefore, IN leaching is dependent on both land use and runoff. Moreover, the runoff plays a dominant role on IN load. Based on the residuals calculated from simulations of discharge, IN concentration and IN load, Figures 3.1-3.4, no significant trend was observed. This indicates that the model structure is robust over time and different spatial scales.

Chapter 4: Comparison of step-wise and multi-objective calibration

In this chapter, applicability of HYPE model wasfurther testedat another mesoscale catchment (Weida) that has different physiographical and chemical characteristics compared with Selke. Step-wise (SWC) and multi-objective (MOC) calibrationwere compared on hydrological water quality parameter optimizationof the HYPE model at Selke and Weida catchments. The objectives were to assess the effects of different calibration procedures on integrated hydrological water quality model identification and theresuggestan appropriate model calibration procedure.

4.1 Modeling results from SWC and MOC at Weida catchment

4.1.1 Model parameter calibration results

4.1.1.1 Parameter sensitivity

Parameters' sensitivities reflected by RCS (relative composite sensitivity) values are shown in Table 4.1. In step-wise calibration (SWC), it is found that the most sensitive hydrological parameter is the correlation parameter for precipitation (*pcadd*) with RCS of 0.026 due to dominant effect of precipitation on runoff generation. The second most sensitive parameter is the potential evapotranspiration rate for arable land (*cevp* for arable land with RCS of 0.024). This is because evapotranspiration is the most important water balance component and arable land is the dominant land use in Weida catchment (40%). The third most sensitive parameter is the maximum velocity in the stream channel (*rivvel* with RCS of 0.01), which is used for calculation of streamflow delay in the river system. The remaining hydrological parameters *ttmp* and *cmlt*, *cevp* for forest and grassland, *srrcs* related to respective snow melt, evapotranspiration for the other important land uses (forest and grass land) and recession of surface water have the same sensitivity order. For nitrogen processes, denitrification rate in the soil (*denitr*) is most sensitive with RCS of 0.06 indicating that denitrification in the soil is very important for the nitrogen balance. The second most sensitive parameter is the half depth of humusN pool for grassland (*hnhalf* for grassland with RCS of 0.003) that is employed to calculate humus nitrogen amount at different soil layers. The parameters *denitw* that defines the denitrification rate in the water and half depth of humusN pool for arable land (*hnhalf* for arable land) have similar lower sensitivity. This indicates that denitrification in the river system is less significant than the denitrification in the soilat Weida catchment. In multi-objective calibration (MOC), the most sensitive parameter is the potential evapotranspiration rate for arable land (*cevp*for arable land with RCS of 0.048). The second most sensitive parameter is the denitrification rate in the soil (*denitr* with RCS of 0.03). Parameters *pcadd* (correlation parameter for precipitation) and *rivvel* (maximum velocity in the stream channel) are also sensitive with the respective RCS of 0.016 and 0.011 (Table 4.1).

Comparing parameter sensitivity analysis results derived from step-wise calibration and multi-objective calibration, the sensitive hydrological water quality parameters determined from both calibration procedures are identical, although the parameter sensitivity ranks are slightly different. This indicates that the hydrological and nitrogen processes of HYPE model are highly identifiable. Trade-off between hydrological and nitrogen processes is noted when multi-objective calibration is applied, which is reflected by lower (higher) RCS values of nitrogen-process (hydrological) parameters derived from MOC than SWC (Table 4.1). Therefore, hydrological processes dominate nitrogen leaching processes. This is consistent with former studies (e.g. Basu *et al.*, 2010; Shrestha *et al.*, 2013).

Table 4.1 Hydrological and nitrogen-process parameters in the HYPE model chosen for calibration at Weida catchment and corresponding calibration results of step-wise calibration (SWC) and multi-objective (MOC) calibration. Shaded rows are the parameters related to hydrological processes, while the remaining are parameters related to nitrogen processes.

Parameter	Physical meaning	Initial value	Initial range	RCS[a] SWC[b]	MOC[c]	OPV[d] SWC	MOC	95% confidence limits SWC	MOC
tmp	Threshold temperature for snow melt and evapotranspiration (°C)								
Arable land		0.044	0.01-10.0	3.98×10^{-3}	4.85×10^{-3}	1.093	0.882	0.543-1.643	0.528-1.236
Forest		1.90	0.01-10.0	7.27×10^{-3}	6.37×10^{-3}	1.728	2.350	0.952-2.504	1.756-2.944
Grassland		1.24	0.01-10.0	4.89×10^{-3}	2.11×10^{-3}	1.546	0.947	0.820-2.272	0.281-1.612
cmlt	Snow melting parameter (mm d^{-1} °C^{-1})								
Arable land		0.55	0.1-10.0	3.52×10^{-3}	4.61×10^{-3}	2.523	0.516	0.977-4.070	0.376-0.656
Forest		4.0	0.1-10.0	3.22×10^{-3}	1.71×10^{-3}	3.60	4.776	1.260-5.941	1.011-8.541
Grassland		1.54	0.1-20.0	3.37×10^{-3}	1.09×10^{-3}	1.757	2.373	0.601-2.912	0.198-4.548
cevp	Potential evapotranspiration rate (mm d^{-1} °C^{-1})								
Arable land		0.13	0.01-10.0	2.37×10^{-2}	4.84×10^{-2}	0.127	0.104	0.117-0.138	0.100-0.107
Forest		1.59	0.01-10.0	7.33×10^{-3}	6.74×10^{-3}	1.741	1.630	1.234-2.242	1.333-1.928
Grassland		0.94	0.01-10.0	7.94×10^{-3}	9.29×10^{-3}	0.963	0.930	0.771-1.155	0.797-1.064
srrcs	Recession coefficient for surface runoff (fraction) (-)								
Arable land		0.085	0.01-1.0	7.60×10^{-3}	4.60×10^{-3}	0.182	0.967	0.160-0.203	0.690-1.243
Forest		0.5	0.01-1.0	1.65×10^{-3}	1.16×10^{-3}	1.00	0.991	0.405-1.595	0.080-1.903
Grassland		0.42	0.01-1.0	2.05×10^{-3}	4.91×10^{-3}	1.00	0.253	0.528-1.472	0.153-0.352
rrvel	Maximum velocity in the stream channel (m s^{-1})	0.1	0.01-10.0	1.01×10^{-2}	1.07×10^{-2}	0.108	0.056	0.098-0.119	0.051-0.061
epotdist	Decrease of evapotranspiration with soil depth (m^{-1})	6.0	0.01-1.0	1.81×10^{-3}	5.18×10^{-3}	6.466	5.417	3.667-9.265	4.382-6.451
pcadd	Correlation parameter for precipitation (-)	0.37	0.01-1.0	2.59×10^{-2}	1.61×10^{-2}	0.479	0.288	0.427-0.531	0.257-0.319
denitr	Denitrification rate in soil (d^{-1})	0.07	0.0001-1.0	6.10×10^{-2}	3.04×10^{-2}	0.062	0.071	0.056-0.068	0.064-0.077
deniw	Parameter for the denitrification in water (kg m^{-2} d^{-1})	0.0001	1-150	3.54×10^{-6}	1.70×10^{-6}	1×10^{-6}	1×10^{-6}	-7.49$\times 10^{-6}$ - 7.51$\times 10^{-6}$	-5.57$\times 10^{-6}$ - 5.59$\times 10^{-6}$
hnhalf	Half depth for humusN pool (m)								
Arable land		0.15	0.01-1.0	3.74×10^{-6}	6.26×10^{-8}	0.011	0.01	-18.31-18.33	-125.9-125.9
Grassland		0.5	0.01-1.0	3.09×10^{-3}	9.54×10^{-4}	0.557	1.0	-0.847-1.961	-1.247-3.247

[a]Relative Composite Sensitivity, [b]Step-Wise Calibration, [c]Multi-Objective Calibration, [d]Optimized Parameter Value.

4.1.1.2 Parameter optimized values and posterior uncertainty

From Table 4.1, it is found that the optimized values (OPV)of most sensitive hydrological water quality parameters estimated from step-wise calibration and multi-objective calibration are similar (e.g. *cevp* and*denitr*). This indicates again that hydrological water quality parameters of HYPE model are highly identifiable. The differences between the optimized values derived from MOC and SWC are attributed to the trade-off between hydrological and nitrogen processes under multi-objective calibration scheme and low parameter sensitivity. Considering parameter posterior uncertainty, uncertainty ranges of most sensitive hydrological and nitrogen-process parameters estimated from MOC are smaller (e.g. *cevp*, *rivvel* and *pcadd*) or similar (e.g. *denitr*) compared with that derived from SWC. Thus, multi-objective calibration enables to better identify hydrological water quality parameters than step-wise calibration by decreasing parameter posterior uncertainty range. Based on comparison of parameter calibration results (e.g. parameter sensitivity and posterior uncertainty) derived from SWC and MOC, multi-objective calibration is concluded to be more appropriate for parameter optimization in integrated hydrological water quality modeling.

4.1.2 Hydrological simulation

As the model performance derived using parameter sets optimized from SWC and MOC are similar in terms of visual representations of discharge and IN dynamics, only the graphics for simulations of streamflow, IN concentration and loadsusing parameter set estimated from multi-objective calibration are shown. The discharge simulation results at the catchment outlet (gauging station Laewitz) for both calibration (1998-2000) and validation (2001-2003) modesare presented in Figure 4.1. The statistical model performances on discharge and IN (daily loads and monthly loads) simulations using estimated parameter sets from SWC and MOC are given in Table 4.2. Hydrographs are represented reasonably well during both calibration and validation periods. Some peak flows are underestimated, which were also reported in former hydrological water quality modeling studies in Weida catchment attributed to uncertainty in input data and model (Balin *et al.*, 2010; Hesser *et al.*, 2010; Shrestha *et al.*, 2007 and 2013).Peak streamflows occur during high intensity rainfall events. However, the HYPE model simulates discharge at daily time step using daily precipitation input. As a result, the model cannot predict the peak flows that occur in short termprecisely because precipitation intensity is averaged over 24-h. In application of WaSIM-ETH for discharge simulation at Weida catchment, Balin *et al.* (2010) used measurements from sixteen rainfall-gauging stations and three climate stations inside/around the Weida catchment to prepare meteorological and climate input data.This may improve discharge prediction accuracy since climate variability can be better covered when more monitoring stations are used. Some recession phases are not well captured (Figure 4.1). This is probably caused by theimproper description of subsurface flow processes in the HYPE model.

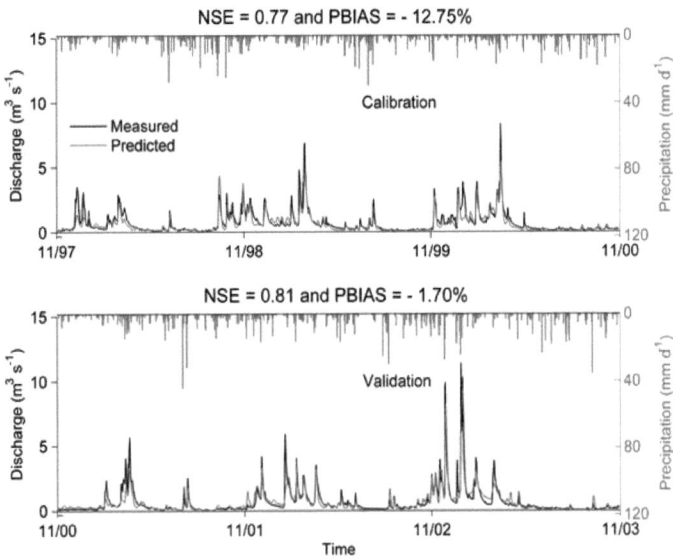

Figure 4.1 Discharge simulation results for both calibration (1998-2000) and validation (2001-2003) modes at the catchment outlet (gauging station Laewitz) using parameter set optimizedfrom multi-objective calibration procedure (MOC).

The statistical performance of the discharge simulations under both SWC and MOC indicate a good agreement between simulated and observed discharges (Table 4.2). According to the watershed model evaluation guidelines specified in Moriasi et al. (2007), HYPE model simulated streamflow good in both calibration and validation modes using parameter sets optimized from step-wise calibration and multi-objective calibration. Comparing model performance derived from SWC and MOC, it is noticed that model performance using parameter set calibrated from SWC is better than that derived from MOC during calibration period in terms of all the criteria (R^2, NSE, PBIAS, and RSR). During validation period, the model performance using parameter set calibrated from SWC is only slightly better than that derived using the parameter set calibrated under MOC in terms of NSE. Model performances are more stable between calibration and validation periods using parameter set calibrated from MOC than that derived using parameter set calibrated from SWC (Table 4.2). Therefore, the optimized parameter setusing MOC is more reasonable and robust than that estimated from SWC. This indicates that calibrating hydrological and water quality parameters simultaneously using both relevant measured variables can improve hydrological parameter identification and increase prediction accuracy.

Table 4.2 Statistical model performances on simulations of discharge, IN concentration, daily IN loads and monthly IN loads for both calibration and validation modes at the catchment outlet (gauging station Laewitz) using the parameter sets optimized from SWC and MOC.

Variable	Criterion	Calibration (1998-2000)		Validation (2001-2003)	
		SWC	MOC	SWC	MOC
Discharge	R^2	0.87	0.79	0.82	0.81
	NSE	0.87	0.77	0.78	0.81
	PBIAS(%)	-0.79	-12.75	11.55	-1.70
	RSR	0.37	0.48	0.47	0.44
IN concentration	R^2	0.15	0.22	0.03	0.08
	$MAE(mgL^{-1})$	3.28	2.86	3.38	3.09
Daily IN load	R^2	0.76	0.75	0.80	0.79
	NSE	0.74	0.71	0.73	0.71
	PBIAS(%)	-14.59	-20.57	-17.47	-19.14
	RSR	0.51	0.54	0.52	0.54
Monthly IN load	R^2	0.80	0.85	0.91	0.89
	NSE	0.77	0.76	0.80	0.77
	PBIAS (%)	-14.60	-20.59	-20.05	-21.27
	RSR	0.48	0.49	0.44	0.48

4.1.3 IN simulation

4.1.3.1 IN concentration and daily IN load

The simulated and observed IN concentrations for both calibration and validation modes at the catchment outlet (gauging station Laewitz) using the parameter set optimized from multi-objective calibration (MOC) are shown in Figure 4.2. Seasonal dynamics of IN concentrations are captured, which are characterized by rapid increase of IN concentrations from autumn to spring and decrease of IN concentrations from spring to autumn. This is the results of combined hydrological and biogeochemical effects (Birgande*t al.*, 2007; Hesser *et al.*, 2010; Shrestha *et al.*, 2007 and 2013; Wagenschein and Rode, 2008). IN is mainly transported through interflow. In winter, the share of interflow increases attributed to precipitation and snowmelt. In summer, runoff is dominated by slow groundwater flow. The longer water residence time in summer results in higher denitrification comparing with winter. Higher temperature in summer also causes higher denitrification and biotic uptake than in winter (Shrestha *et al.*, 2013).

Underestimation of IN concentrations during some winter periods (e.g. winters of 1997-1998 and 2000-2001) is observed, which is mainly caused by the underestimation of high discharge. In addition, the point sources such as waste water effluent not connected to sewage systems in the catchment and/or runoff from agricultural drainage from nearby stream zones also contributes to the IN concentration underestimation(Hesser et al., 2010; Shrestha et al., 2007); noises/measurement errors in IN observations may also deteriorate model performance. There is a clear relationship of the goodness-of-fit between the hydrological simulations and IN concentration simulations, which indicates the importance of good simulations of hydrological processes for satisfactory IN transport simulations (Figures 4.1 and 4.2).

According to Table 4.2, model performance on IN concentration simulations using parameter set optimized from MOC are higher and more robust than using parameters estimated from SWC shown by R^2 and MAE. The mean of measured IN concentrations for the calibration period and validation period are 8.99 mg L^{-1} and 8.53 mg L^{-1}, respectively. Corresponding mean of simulated IN concentrations under SWC (MOC) for calibration and validation periods are 7.94 mg L^{-1} (8.35 mg L^{-1}) and 7.36 mg L^{-1} (7.98 mg L^{-1}), respectively. Therefore, model predicts mean IN concentration more accurately under multi-objective calibration comparing with step-wise calibration.

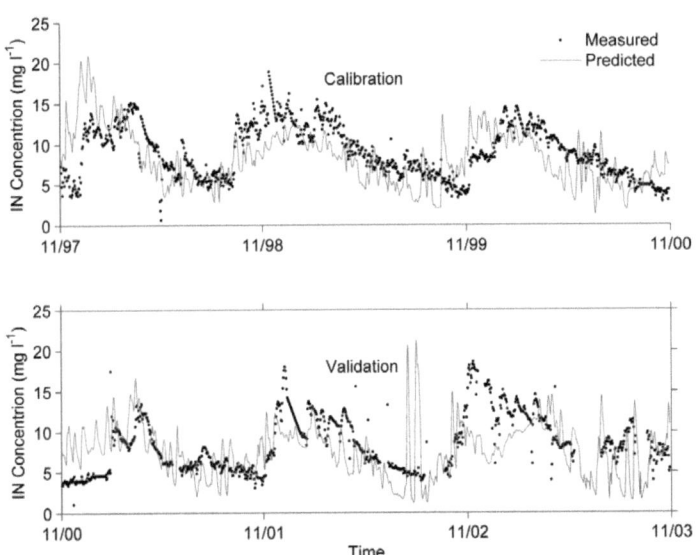

Figure 4.2 Simulated and observed IN concentrations for both calibration (1998-2000) and validation (2001-2003) modes at the catchment outlet (gauging station Laewitz) using parameter set optimized using multi-objective calibration (MOC).

Simulated and measured daily IN loads for calibration and validation modes at the catchment outlet (gauging station Laewitz) using the parameter set optimized from multi-objective calibration (MOC) are shown in Figure 4.3. The statistical model performance on daily IN loads simulations using parameter sets optimized from SWC and MOC are given in Table 4.2. The dynamics of daily IN loads are represented relatively well during both calibration and validation periods. Daily IN loads show similar seasonal dynamics as discharge and streamwater IN concentration, which are characterized by high loads in winter and lower values in summer. Some peak daily IN loads during winter-spring seasons (e.g. 1998-1999 and 2002) were underestimated, which are mainly attributed to underestimation of peak discharges and consequently underestimation of high streamwater IN concentrations (Figures 4.1-4.3). This is also the reason for the overall underestimation of daily IN loads in both calibration and validation periods indicated by PBIAS of around – 20%. Based on comparison of statistical model performances on daily IN loads simulations using parameter sets optimized from SWC and MOC given in Table 4.2, model performance under MOC are more robust between calibration and validation modes than SWC, although the latter are slightly better than the former for the same period.

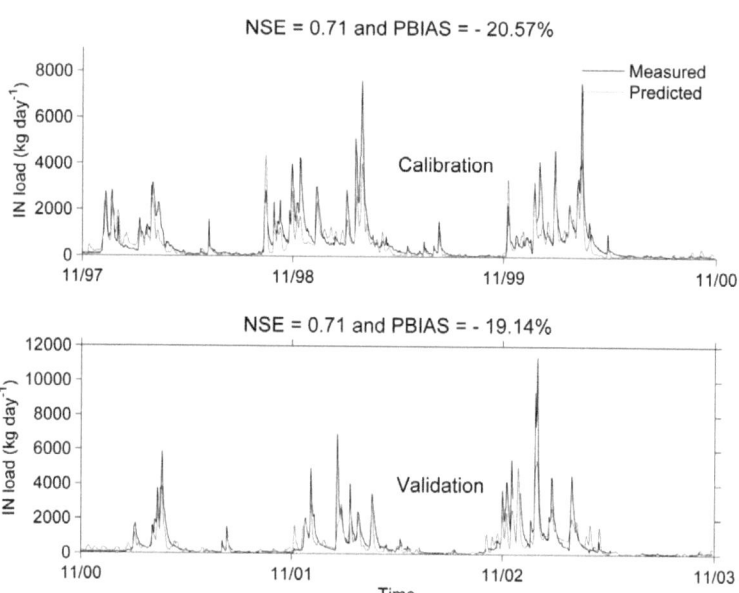

Figure 4.3 Simulated and observed daily IN load for both calibration (1998-2000) and validation (2001-2003) modes at the catchment outlet (gauging station Laewitz) using parameter set optimized from multi-objective calibration (MOC).

4.1.3.2 Monthly and annual IN loads simulation

To test the model capability in predicting IN loads at larger time intervals than daily, monthly IN loads were calculated at the catchment outlet (gauging station Laewitz) for both calibration (1998-2000) and validation (2001-2003) periods. The simulation results using the parameter set estimated from MOCarepresented in Figure 4.4. The statistical model performance on monthly IN loads simulation under SWC and MOC are given in Table 4.2.The method to calculate the simulated and measured monthly IN loads has been explained in Chapter 2 (Section 2.5.2.1). Similar as dynamics of daily IN loads, monthly IN loads show obvious seasonal dynamics, which is characterized by high loads during high flow periods in winter-spring months and lower values during low flow conditions in summer-autumn months. Due to underestimation of peak discharges and IN concentrations in winter periods, monthly IN loads were overall underestimated for both calibration and validation modes indicated by PBIAS values (Table 4.2).Comparing model performance on monthly IN loads simulation using parameter sets calibrated from MOC and SWC, model performance under SWC is noticed to be better than MOC for the same period. However, parameter values optimized from MOC are more robust than that estimated from SWC, which is reflected by the more stable model performance (R2, NSE, PBIAS and RSR) between calibration and validation periods derived from MOC.

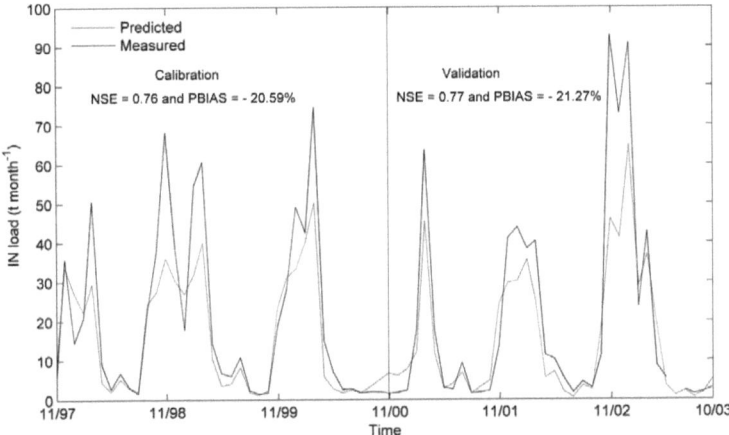

Figure 4.4 Simulated and observed monthly IN load at the catchment outlet (gauging station Laewitz) for both calibration (1998-2000) and validation (2001-2003) modes using parameter set optimized using multi-objective calibration (MOC).

The mean simulated and measured monthly IN loads at the catchment outlet (gauging station Laewitz) using the parameter set estimated from multi-objective calibration (MOC) for both calibration (1998-2000) and validation (2001-2003) modes were calculated and presented in Figure 4.5. Mean monthly IN loads show clear seasonal dynamics, which are characterized by high loads from November to March and low loads between April and October. Corresponding to model performance on IN loads simulations at daily and monthly intervals, underestimation of mean monthly IN loads is noted during winter-spring months characterized by high streamflow and IN concentrations. This indicates the importance of correct discharge and IN concentration simulations for IN load calculation.

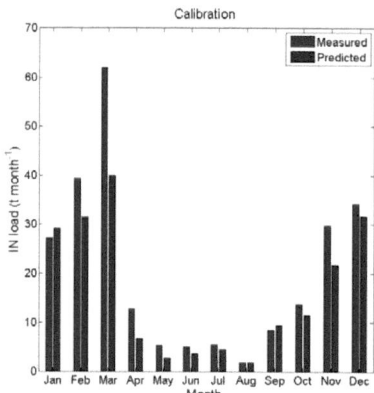 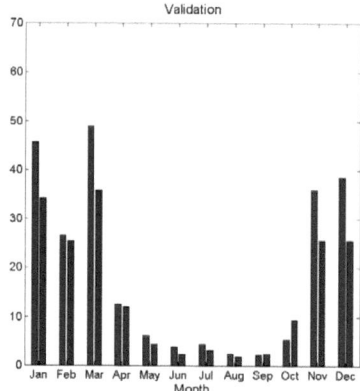

Figure 4.5 Averages of simulated and measured monthly IN load for both calibration (1998-2000) and validation (2001-2003) modes at the catchment outlet (gauging station Laewitz) using the parameter set optimized using multi-objective calibration (MOC).

To investigate the spatial variability of IN losses attributed to heterogeneity in climate, land use and soil, mean annual IN losses contributed by each sub-basin during calibration period (1998-2000) and validation period (2001-2003) were calculated using the parameter set optimized from multi-objective calibration (MOC) and shown in Figure 4.6. Similar patterns of spatially distributed IN loads were found for calibration and validation modes. Also, the levels of annual IN losses from each sub-basin are similar during calibration and validation periods. Therefore, only the spatial variability of area weighted annual IN losses in calibration period is discussed below.

Sub-basin 22 has the lowest IN losses, which is 1.61kg $ha^{-1} y^{-1}$ corresponding to the low share of arable land (9%) and low runoff of 13 mm y^{-1}. Sub-basins 30, 26, 13, 15, 17, 14 and 3 have similar low levels of IN losses as sub-basin 22, ranging 2.60-6.41kg $ha^{-1} y^{-1}$. All these sub-basins are located in the boundary of the catchment and have low runoff (<60 mm y^{-1}). Sub-basins 27, 6 and 20 have IN leaching loads between 10.47 and 11.01kg $ha^{-1} y^{-1}$ corresponding to arable land share of 46-55%. Sub-basin 23 contributes IN losses of 10.13 kg $ha^{-1} y^{-1}$ attributed to high runoff (108 mmy^{-1}), despite of low arable land share (2%). Sub-basin 8 has IN losses of 16.86 kg $ha^{-1} y^{-1}$ corresponding to arable land share of 74% and annual runoff of 140 mm. Sub-basins 32, 29, 25, 24, 19, 18, 12, 16, 11, 9, 10 and 4 have high IN losses between 17.62 and 21.38 kg $ha^{-1} y^{-1}$, attributed to high share of arable land (e.g. sub-basin 32) and/or high runoff (e.g. sub-basin 19 with annual runoff of 230 mm).Sub-basin 5 has the highest IN losses of 26.19 kg $ha^{-1} y^{-1}$ attributed to high runoff of 333 mmy^{-1}. Based on the above analysis onspatial variability of IN losses, it indicates that IN leaching are dependent on both land use (share of arable land) and runoff, while, runoff plays a dominant role.

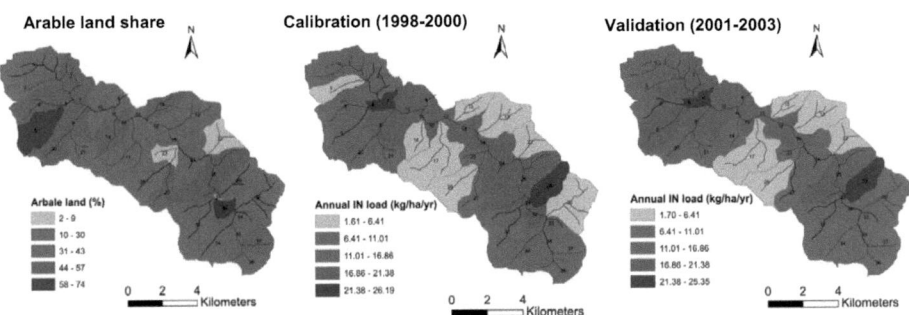

Figure 4.6 Average simulated annual IN losses by each sub-basin of Weida catchment for calibration (1998-2000) and validation (2001-2003) periods (middle and right) using parameter set optimized from multi-objective calibration (MOC); the share of arable land in each sub-basin is also presented to show relationship between land use and IN losses.

4.2 Modeling results from SWC and MOC at Selke catchment

4.2.1 Model parameter calibration results

Hydrological water quality parameter calibration results derived from step-wise calibration (SWC) and multi-objective calibration (MOC) are given in Table 4.3. Considering parameter sensitivity, all five calibrated hydrological parameters have similar magnitudes of relative composite sensitivity (RCS) under both SWC and MOC. The most and second most sensitive hydrological parameters determined from SWC and MOC are identical, which are the maximum stream flow velocity (*rivvel*) and potential evapotranspiration rate for agricultural land (*cevp*). The sensitivity ranks for the remaining hydrological parameters are different under SWC and MOC.For nitrogen processes, the top three most sensitive parameters are the Fraction of nutrient uptake in the uppermost soil layer for agricultural land and mixed forest (uptsoil1) and denitrification rate in soil (*denitr*), which is consistent between SWC and MOC.

Comparing the best estimated parameter values from SWC and MOC, there is no significant difference between the two optimized parameter sets. This indicates that hydrological water quality parameters of HYPE model are highly identifiable. Concerning parameter posterior uncertainty, the 95% parameter uncertainty ranges derived from MOC are much narrower than that estimated from SWC, especially for nitrogen-process parameters. This indicates that calibrating hydrological and nitrogen-process parameters simultaneously enables to improve identification of nitrogen-process parameters comparing with step-wise calibration because (i) both processes are connected; (ii) nitrogen transport and transformation are highly dependent on hydrological variables (e.g. flow components and soil moisture).

4.2.2 Model simulation results

Model statistical performance on simulations of discharge, IN concentration, daily and monthly IN loads at three gauging stations (Silberhuette, Meisdorf and Hausneindorf) in Selke catchment using parameter sets optimized from step-wise calibration (SWC) and multi-objective calibration (MOC) are given in Table 4.4. There is no significant difference between model performances derived from SWC and MOC in simulations of discharge, IN concentration and loads. In discharge simulation, model performances derived from MOC are slightly better than SWC in both calibration and validation periods at gauging stations Silberhuette and Meisdorf in terms of PBIAS. For IN concentration simulation, model performance at catchment outlet (gauging station Hausneindorf) derived from MOC is slightly better than SWC. The high similarity between model performances in simulations of discharge, IN concentration and loads derived from SWC and MOC can be explained that hydrological water quality parameters estimated from both calibration procedures are nearly identical attributed to proper definition of initial values and ranges.This also indicates the identical optimized parameter set estimated from step-wise calibration and multi-objective calibration are global optimal values.

Table 4.3 Hydrological and nitrogen-process parameters in the HYPE model chosen for calibration at Selke catchment and corresponding calibration results of step-wise calibration (SWC) and multi-objective (MOC) calibration. Shaded rows are the parameters related to hydrological processes, while the rest are parameters related to nitrogen processes.

Parameter	Physical meaning	Initial value	Initial range	RCS^a SWC^b	MOC^c	OPV^d SWC	MOC	95% confidence limits SWC	MOC
cevp	Potential evapotranspiration rate ($mm\ d^{-1}\ °C^{-1}$)								
Agricultural land		0.22	0.01-1.0	5.25×10^{-3}	5.02×10^{-3}	0.230	0.230	0.212-0.249	0.218-0.241
Coniferous forest		0.16	0.01-1.0	4.34×10^{-3}	4.39×10^{-3}	0.176	0.170	0.167-0.184	0.162-0.179
Mixed forest		0.12	0.01-1.0	3.41×10^{-3}	4.53×10^{-3}	0.120	0.116	0.114-0.125	0.110-0.122
rivvel	Maximum velocity in the stream channel ($m\ s^{-1}$)	0.20	0.001-1.0	6.01×10^{-3}	5.83×10^{-3}	0.199	0.202	0.195-0.203	0.197-0.206
epotdist	Decrease of evapotranspiration with soil depth (m^{-1})	6.646	1.0-10.0	3.96×10^{-3}	3.90×10^{-3}	6.659	6.407	5.986-7.331	5.982-6.831
denitr	Denitrification rate in soil (d^{-1})	0.01	0.001-0.1	0.065	2.56×10^{-3}	0.0227	0.0222	0.017-0.028	0.019-0.025
denitw	Parameter for the denitrification in water ($kg\ m^{-2}\ d^{-1}$)	0.001	0.0001-0.01	1.75×10^{-3}	7.16×10^{-5}	0.0001	0.0001	-0.0009-0.0011	-0.0003-0.0005
wprod	Production/decay of N in water ($kg\ m^{-3}\ d^{-1}$)	0.001	0.0001-0.01	0.0103	4.18×10^{-4}	0.0075	0.0076	0.0004-0.0145	0.0047-0.011
uptsoil1	Fraction of nutrient uptake in the uppermost soil layer (-)								
Agricultural land		0.5	0.01-1.0	0.269	0.0106	1.0	1.0	0.945-1.055	0.976-1.024
Coniferous forest		0.5	0.01-1.0	1.99×10^{-4}	2.08×10^{-3}	0.901	1.0	-40.99-42.80	0.809-1.191
Mixed forest		0.5	0.01-1.0	0.126	3.78×10^{-3}	0.963	0.961	0.848-1.077	0.879-1.043

[a] Relative Composite Sensitivity, [b] Step-Wise Calibration, [c] Multi-Objective Calibration, [d] Optimized Parameter Value.

Table 4.4 Model statistical performance on simulations of discharge, IN concentration, daily and monthly IN loads at the gauging stations Silberhuette, Meisdorf and Hausneindorf during the calibration (1994-1999) and validation (1999-2004) periods using the parameter sets optimized from step-wise calibration (SWC) and multi-objective calibration (MOC).

Variable	Criterion	Calibration						Validation					
		Silberhuette		Meisdorf		Hausneindorf		Silberhuette		Meisdorf		Hausneindorf	
		SWC	MOC	SWC	MOC	SWC	MOC	SWC	MOC	SWC	MOC	SWC	MOC
Discharge	R^2	0.89	0.88	0.88	0.88	0.86	0.86	0.92	0.92	0.90	0.90	0.88	0.88
	NSE	0.88	0.88	0.88	0.88	0.86	0.86	0.91	0.91	0.90	0.90	0.86	0.86
	PBIAS	-5.77	-4.96	-5.19	-4.06	1.96	2.41	-11.2	-10.36	-2.19	-1.03	13.49	14.0
	RSR	0.34	0.34	0.35	0.35	0.38	0.38	0.31	0.31	0.32	0.32	0.37	0.37
IN con-centration	R^2	0.72	0.72	0.58	0.58	0.08	0.09	0.43	0.43	0.33	0.32	0.05	0.06
	MAE $(mg\ L^{-1})$	0.47	0.48	0.50	0.51	0.87	0.86	0.68	0.67	0.50	0.50	1.0	1.0
Daily IN load	R^2	0.92	0.92	0.82	0.82	0.72	0.71	0.83	0.83	0.91	0.91	0.82	0.82
	NSE	0.88	0.88	0.80	0.80	0.70	0.69	0.83	0.83	0.89	0.89	0.43	0.43
	PBIAS	-10.81	-10.80	-14.78	-14.73	5.54	5.24	1.27	1.37	8.72	8.84	41.51	40.90
	RSR	0.34	0.34	0.45	0.45	0.55	0.55	0.41	0.41	0.34	0.34	0.75	0.75
Monthly IN load	R^2	0.91	0.91	0.86	0.86	0.84	0.83	0.86	0.86	0.90	0.90	0.87	0.87
	NSE	0.85	0.85	0.77	0.77	0.80	0.79	0.86	0.86	0.91	0.91	0.68	0.68
	PBIAS	-14.90	-14.98	-21.01	-21.06	-5.75	-6.29	0.99	1.05	-0.44	-0.33	34.28	33.88
	RSR	0.39	0.39	0.48	0.48	0.45	0.46	0.38	0.38	0.31	0.31	0.57	0.57

Chapter 5: Effects of spatial and temporal resolution of calibration data on model identification

As discussed in Chapter 1 (section 1.2.6), spatial and temporal resolutions of calibration data (e.g. discharge andstreamwater nutrient concentrations) may impact model parameter identification, model performance and prediction accuracy. In this chapter, PEST and DREAM$_{(ZS)}$ were combined with the HYPE model to implement and compare parameter automatic calibration and uncertainty analysis using calibration data of different spatial and temporal resolutions. The procedures of this study have been explained in Chapter 2 (Section 2.8). The objectives were to (i) investigate the influences of spatial and temporal resolutions of streamflow and IN concentration observations on hydrological water quality parameter identification (e.g. parameter posterior uncertainty), model performance, robustness of the optimized parameters and prediction accuracy; (ii) to evaluate the efficiency and reasonableness of PEST and DREAM$_{(ZS)}$ on hydrological water quality model calibration and uncertainty analysis.

5.1 Effects of spatial resolution of calibration data

5.1.1 Discharge simulation

5.1.1.1 Calibration and predictive analysis using PEST

In this part, PEST was combined with the HYPE model to implement hydrological parameter automatic calibration and predictive analysis. Through comparing parameter calibration results, model performance and predictive uncertainty derived using discharge measurements from different number of discharge gauging stations (i.e. single-site vs. multi-site calibration), information content of multi-sites treamflow monitoring strategy was assessed.

Parameter identification

Following the procedure of parameterization described in Chapter 2 (Section 2.3.4), five sensitive hydrological parameters were chosen for calibration and predictive analysis in this part. Their Physical meanings, initial values and intervals have been given in Table 4.3.

The model setup followed the same procedure described in Chapter 2 (Section 2.5.1.1). Discharge measured during the period 1994-1999 was used for model calibration and one year (1993-1994) was used for model warming up, which was excluded for model evaluation. Four different calibration-predictive analysis modes were implemented and compared, which involve discharge measurements from different number of gauging stations. They were categorized as (i) Mode 1 (single-site calibration and predictive analysis that uses discharge observations only from catchment outlet (gauging station Hausenindorf)); (ii) Mode 2 (two-site calibration and predictive analysis that uses discharge observations from catchment outlet (gauging station Hausneindorf) and headwater (gauging station Silberhuette)); (iii) Mode 3 (two-site calibration and predictive analysis that uses discharge observations from catchment outlet (gauging station Hausneindorf) and middle part of the river (gauging station Meisdorf) and (iv) Mode 4 (three-site calibration and predictive analysis that uses discharge observations from catchment outlet (gauging station Hausneindorf) and both two internal sites (gauging stations Silberhuette and Meisdorf)). In each mode, predictive analysis was conducted after corresponding calibration mode according to the procedure of implementing predictive analysis using PEST, which has been described in detail in Chapter 2 (Section 2.4.1.3) and Appendix 2C. The spatial and temporal robustness of parameter sets optimized from calibration modes 1-4were compared. The model performances using parameter sets estimated from minimum and maximum predictive analysis of mode 1-4 were compared with that using corresponding optimized parameter set.

The normalized 95% posterior confidence intervals of each calibrated parameter estimated from different calibration modes are shown in Figure 5.1. The lower and upper limit of each parameter is divided to its optimized value in order to compare the uncertainty margins of all calibrated parameters. To better compare parameter uncertainty derived from different calibration modes, the normalized posterior parameter uncertainty ranges derived from calibration Modes 1- 4 are shown in one plot following the general decreasing parameter relative composite sensitivity. It needs to be mentioned that there are small differences in the rank of parameters' relative composite sensitivity estimated from different calibration modes due to spatial variability of catchment characteristics and hydrological regimes, which influences the sensitivity of a certain parameter. It is found that potential evapotranspiration coefficient for agricultural land (*cevp2*) is most sensitive under all different calibration modes, which can be explained by the fact that evapotranspiration is the most important component of water balance and the Selke catchment is dominated by agricultural land. The same ranks of parameters' relative composite sensitivity are found for calibration Mode 1 and Mode 3. Similar ranks are found for calibration Mode 2 and Mode 3; the small difference is reflected in the rank of relative composite sensitivity for *cevp7* and *cevp8* characterized by higher sensitivity of *cevp7* than *cevp8* in calibration Mode 2 and following opposite rank for Mode 3. This is attributed to the higher share of coniferous forest and lower share of mixed forest in sub-basin Silberhuette compared to sub-basin Meisdorf. For the calibration Mode 4, the rank of parameters' relative sensitivity is slightly different from that derived from all other three modes (Modes 1-3); the higher order of relative composite sensitivity of the parameter *rivvel* (Maximum velocity in the stream channel) compared with that estimated from the other three modes verifies that river flow velocity issensitive for calculation of regional river flow delay. The parameter sensitivity analysis gives important indication of the important hydrological processes; the rank of parameter sensitivity may change when observations from internal sites are involved in parameter calibration, especially if the catchment shows high spatial heterogeneity (Van Griensven and Bauwens, 2003).

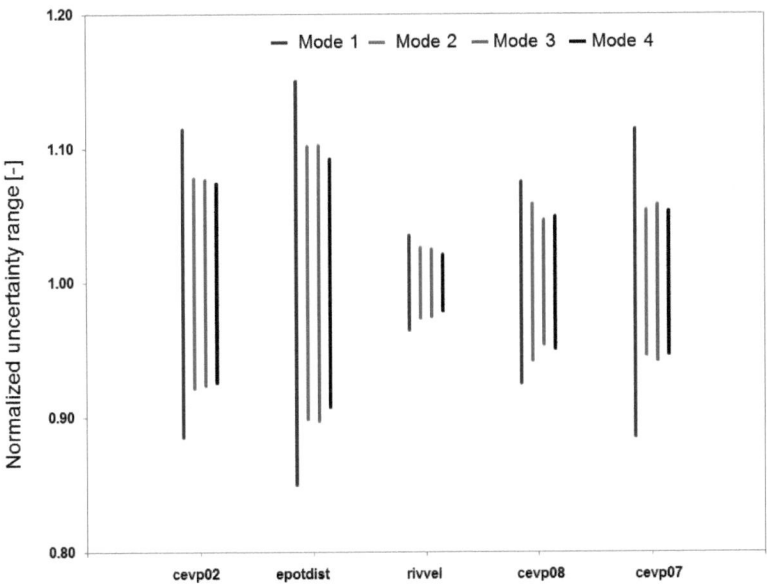

Figure 5.1 Normalized 95% uncertainty margin of each calibrated hydrological parameter estimated from four calibration modes (Mode 1- Mode 4) following general decreasing parameter relative composite sensitivity.

From Figure 5.1, it is found that the largest uncertainty range for each parameter was estimated from single-site calibration (Mode 1). A similar uncertainty range for each parameter was derived when the model was calibrated under two-site calibration modes (Modes 2 and 3). The narrowest uncertainty range for each parameter was found when three-site calibration was applied (Mode 4). The differences in parameters' uncertainty ranges estimated from different calibration modes indicate that parameter posterior uncertainty range decreases and parameter identification improves when multi-site calibration is applied. There are no large differences between parameter uncertainty ranges estimated from two-site calibration

(Modes 2 and 3) and three-site calibration (Mode 4), which indicates that (i) adding discharge observations from either internal site (Silberhuette/Meisdorf) for parameter calibration has similar effect on decreasing parameter uncertainty; (ii) adding discharge observations from two internal sites rather than one does not obviously further decrease parameter uncertainty. This can be explained by the fact that the two internal gauged sub-basins (Silberhuette and Meisdorf) have similar catchment characteristics (climate patterns, topography, land use and soil type). Therefore, the discharge observations from these two internal gauging stations are highly correlated (correlation coefficient R=0.97). As a result, adding discharge observations from two internal sites rather than one does not further increase information content.

The optimized parameter sets and estimated parameter sets from calibration-predictive analysis (minimum and maximum predictive analysis) of Mode 1- Mode 4 are given in Table 5.1. In each mode the differences between calibrated parameter set and parameter set estimated from minimum and maximum predictive analysis were investigated using t-test based on the theory of two sample t-test with unequal variances (Bahremand and De Smedt, 2010; Helsel and Hirsch, 1992). It is found that under each mode the parameter sets estimated from minimum and maximum predictive analysis are statistically identical as the corresponding optimized parameter setbecause their 95% confidence intervals overlap each other, although in Modes 1- 3 the calibrated *rivvel*is statistically different from the value estimated from minimum predictive analysis attributed to parameter interaction. This indicates that the hydrological parameters of the HYPE model are highly identifiable; the model structure is physically sound and robust.

Model performance

The statistical model performances on discharge simulations during calibration and predictive analysis (1994-1999) and validation (1999-2004) periodsusing the parameter sets estimated from respective calibration and predictive analysis of Modes 1-4 are shown in Table 5.2. The HYPE model had similar simulation results during both calibration and validation periods using parameter sets calibrated from different modes. Therefore, only the discharge simulation results using parameter set optimized from single-site calibration (Mode 1) are shown. The simulated and measured discharge together with daily precipitation for calibration and temporal and spatial validation are presented in Figure 5.2. The hydrographs are represented very well for all calibration and validation modes as well as predictive analysis modes with the lowest NSE of 0.86.Both high flow events and low flows were well captured (Figure 5.2). Using the parameter set calibrated against discharge observations from catchment outlet (Mode 1), the HYPE model had similar model efficiency between calibration (NSE = 0.86) and spatial validation at internal sub-basins that show high variability in terms of topography, land use and soil type (NSE = 0.88 at Silberhuette and Meisdorf). The catchment characteristics heterogeneityresults in spatial variation of hydrological regimes characterized by quick subsurface flow dominated runoff in the upper-stream mountain areas and slow surface flow dominated runoff in the lowland areas. The difference in water balance between catchment outlet and internal sites (PBIAS $\approx$ 0 at Hausneindorf and PBIAS $\approx$ - 6% at Silberhuette and Meisdorf) is mainly caused by different evapotranspiration patterns between upper-stream mountain areas and downstream lowland areas corresponding to spatial variability of climate and land use. Due to a simplified evapotranspiration routine, the HYPE model cannot represent properly the processes driving the spatial variation of evapotranspiration (Strömqvist *et al.*, 2012). The impact of topography on evapotranspiration may need to be considered in future model development.

Table 5.1 The optimized and estimated parameter sets from respective calibration and predictive analysis (Minimum and Maximum prediction modes) under Mode 1- Mode 4.

Parameter	Calibration mode		Minimum prediction mode		Maximum prediction mode		t test result	
	Estimated value (m_x)	Standard deviation (S_x)	Estimated value (m_y)	Standard deviation (S_y)	Estimated value (m_z)	Standard deviation (S_z)		
			Mode 1					
cevp2	0.240	0.0140	0.241	0.0178	0.230	0.0152	$m_x = m_y$	$m_x = m_z$
cevp7	0.164	0.0095	0.173	0.0104	0.161	0.0106	$m_x = m_y$	$m_x = m_z$
cevp8	0.127	0.0049	0.124	0.0068	0.126	0.0066	$m_x = m_y$	$m_x = m_z$
rivvel	0.212	0.0038	0.201	0.0035	0.212	0.0035	$m_x \neq m_y$	$m_x = m_z$
epotdist	6.664	0.5089	6.645	0.5868	6.640	0.5308	$m_x = m_y$	$m_x = m_z$
			Mode 2					
cevp2	0.234	0.0093	0.236	0.0119	0.2273	0.0094	$m_x = m_y$	$m_x = m_z$
cevp7	0.174	0.0048	0.177	0.0064	0.1747	0.0051	$m_x = m_y$	$m_x = m_z$
cevp8	0.123	0.0036	0.124	0.0044	0.1265	0.0037	$m_x = m_y$	$m_x = m_z$
rivvel	0.205	0.0028	0.194	0.0025	0.2097	0.0030	$m_x \neq m_y$	$m_x = m_z$
epotdist	6.670	0.3450	6.645	0.4208	6.6320	0.3504	$m_x = m_y$	$m_x = m_z$
			Mode 3					
cevp2	0.235	0.0092	0.241	0.0129	0.225	0.0088	$m_x = m_y$	$m_x = m_z$
cevp7	0.178	0.0053	0.179	0.0066	0.176	0.0054	$m_x = m_y$	$m_x = m_z$
cevp8	0.118	0.0028	0.120	0.0040	0.125	0.0030	$m_x = m_y$	$m_x = m_z$
rivvel	0.210	0.0027	0.199	0.0025	0.214	0.0028	$m_x \neq m_y$	$m_x = m_z$
epotdist	6.658	0.3455	6.647	0.4036	6.751	0.3305	$m_x = m_y$	$m_x = m_z$
			Mode 4					
cevp2	0.258	0.0098	0.255	0.0088	0.264	0.0063	$m_x = m_y$	$m_x = m_z$
cevp7	0.163	0.0044	0.163	0.0050	0.170	0.0041	$m_x = m_y$	$m_x = m_z$
cevp8	0.124	0.0031	0.121	0.0034	0.122	0.0031	$m_x = m_y$	$m_x = m_z$
rivvel	0.203	0.0021	0.199	0.0021	0.201	0.0021	$m_x = m_y$	$m_x = m_z$
epotdist	7.326	0.3450	6.830	0.3300	8.017	0.2609	$m_x = m_y$	$m_x = m_z$

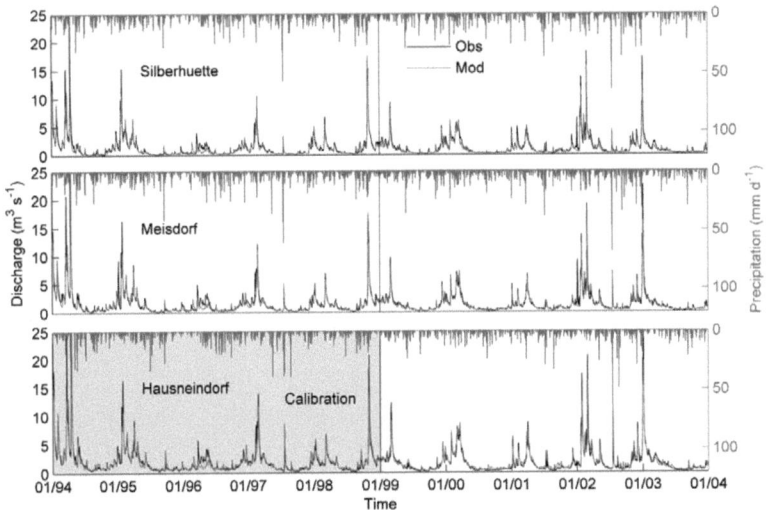

Figure 5.2 Discharge simulation results for calibration, temporal and spatial validation using the parameter set optimized from single-site calibration (Mode 1).

The similar model efficiencies (NSE) across three gauging stations indicate that structure of the HYPE model (i.e. simulated hydrological processes and parameter set) is transferable in space. Temporally it showed slightly higher model efficiencies during validation period than calibration period (NSE = 0.87 vs. NSE = 0.86 at Hausneindorf, NSE = 0.90 vs. NSE = 0.88 at Silberhuette and Meisdorf); larger underestimation at headwater (Silberhuette) during validation mode than calibration mode may be caused by interannual climate variation and uncertainty in input data of precipitation. The discharge overestimation at catchment outlet during validation period is mainly caused by discharge measurement errors due to effects of streamwater extraction. The mismatches between simulated and measured discharge during February - May, 1996 could be caused by streamflow measurement errors. Constant and good model performances between calibration and validation periods indicate that the HYPE model structure is transferable temporally.

Table 5.2 Statistical model performances on discharge simulations during calibration (1994-1999) and validation (1999-2004) periods, minimum and maximum predictive analysis modes (1994-1999) using the parameter sets estimated from respective calibration and predictive analysis of Mode 1- Mode 4.

Gauging station	Calibration 1994-1999		Validation 1999-2004		Minimum prediction 1994-1999		Maximum prediction 1994-1999	
	NSE	PBIAS (%)	NSE	PBIAS (%)	NSE	PBIAS (%)	NSE	PBIAS (%)
Mode 1								
Silberhuette	0.88	- 6.38	0.90	- 11.87	0.88	- 6.93	0.88	- 5.18
Meisdorf	0.88	- 6.29	0.90	- 3.29	0.88	- 6.69	0.88	- 5.06
Hausneindorf	0.86	0.02	0.87	11.93	0.86	- 0.58	0.86	1.96
Mode 2								
Silberhuette	0.88	- 6.25	0.90	- 11.68	0.88	- 7.04	0.88	- 6.65
Meisdorf	0.88	- 5.91	0.90	-2.95	0.88	- 6.86	0.88	- 6.80
Hausneindorf	0.86	0.66	0.87	12.59	0.86	- 0.48	0.86	0.40
Mode 3								
Silberhuette	0.88	- 6.17	0.90	- 11.45	0.88	- 7.07	0.88	- 6.16
Meisdorf	0.88	- 5.44	0.90	-2.39	0.88	- 6.55	0.88	- 6.13
Hausneindorf	0.86	0.85	0.87	12.87	0.86	- 0.55	0.86	1.54
Mode 4								
Silberhuette	0.88	- 5.74	0.91	- 11.03	0.88	- 5.93	0.88	- 5.23
Meisdorf	0.88	- 4.94	0.90	- 1.58	0.88	- 5.03	0.88	- 4.07
Hausneindorf	0.86	1.43	0.86	13.98	0.86	0.50	0.86	3.18

From Table 5.2, it is found that the HYPE model had similar performances for whole simulation periods (calibration period of 1994-1999 and validation period of 1999-2004) at all three gauging stations using parameter sets optimized from different calibration modes (Mode1- Mode 4). This can be explained by the fact that (i) measured discharges from catchment outlet (gauging station Hausneindorf) were included in all four calibration modes, which represent the hydrological responses of the whole Selke catchment to the climate forcing; (ii) all three gauged sub-basins (Silberhuette, Meisdorf and Hausneindorf) are hydrologically connected due to the "nested" nature of the catchment; (iii) the high similarity in observed hydrographs from all three sites indicates that the hydrological processes at different locations of the catchment are highly correlated. In other words, adding discharge observations from internal site(s) for parameters calibration does not increase much information on hydrological processes and therefore cannot greatly improve parameter identification and model performance. It is noted that there are some trade-offs between sub-basins when discharge measurements from internal site(s) were added for parameter calibration (i.e. multi-site calibration). This can be obviously seen by comparing calculated PBIAS of each gauging station using parameter sets optimized from calibration Mode 1 and Mode 4 (Table 5.2). It shows that the discharge underestimation at the internal sites (gauging stations Silberhuette and Meisdorf) decreased and overestimation at the catchment outlet (gauging station Hausneindorf) increased slightly when discharge observations from the two internal sites were added in parameters calibration.

Predictive uncertainty

As described in Chapter 2 (Section 2.4.1.3) and Appendix 2C, predictive analysis using PEST is preceded by parameter estimation in which Φ_{min} is determined. Taking the single-site calibration and predictive analysis (Mode 1) as an example, the minimum objective function Φ_{min} obtained during the calibration is 2365. The value of δ is calculated as 14, which is about 0.59% of the Φ_{min} and therefore Φ_0 was set to 2379. The sole number of the model outputs (named as "key model prediction"), which was considered as the predict group, was chosen as the winter flood occurred on the 14th of April, 1994. The measured discharge of this flood is 56.03 $m^3 s^{-1}$. After calibration, the model is run in prediction mode twice. In other words, once PEST has minimized the objective function to get the best prediction, it is asked to maximize/minimize the target prediction while maintaining the model calibration. The calibrated and predicted minimum and maximum values for this winter flood calculated from four calibration and predictive analysis modes are given in Table 5.3. The model performances on discharge simulations using the parameter sets estimated during minimum and maximum predictive analysis modes for the period 1994-1999 are shown in Table 5.2. The HYPE model had similar performances using the parameter sets estimated from minimum and maximum predictive analysis as with the parameter set optimized from corresponding calibration at each gauging station. It is found that the HYPE model had slightly better performances using the parameter set estimated from maximum predictive analysis than with that estimated from minimum predictive analysis in terms of PBIAS. Similar as the finding derived from calibration procedures, there are small trade-offs between sub-basins when discharge observations from internal site(s) were included for predictive analysis. The similar model performances between model calibration and predictive analysis modes indicate that parameter non-uniqueness does not influence model prediction accuracy significantly.

Table 5.3 The calibrated discharge, minimum and maximum predictions for the winter flood occurred on 14 April 1994 at the catchment outlet (gauging station Hausneindrof), estimated from four different calibration and predictive analysis modes (Mode 1- Mode 4).

Mode	Calibrated (m^3s^{-1})	Minimum prediction (m^3s^{-1})	Maximum prediction (m^3s^{-1})
Mode 1	48.02	46.66	49.51
Mode 2	47.83	46.05	49.24
Mode 3	47.87	46.35	51.10
Mode 4	47.21	46.38	49.78

Figure 5.3 presents observed discharge, simulated discharge with best optimized parameter set and predictive uncertainty ranges shaped by minimum and maximum predictions, which are derived from different calibration and predictive analysis modes (Mode 1- Mode 4) at the catchment outlet (gauging station Hausneindorf) for a two-month period (April-May of 1994). Best simulated discharges lie between the minimum and maximum predictions in all four different calibration and predictive analysis modes. Under each mode, the differences between best simulated discharges, minimum and maximum predictions are small, especially during low flow conditions reflected by narrow prediction uncertainty range (grey area in Figure 5.3). The discharge predictive uncertainty ranges estimated from Mode 2- Mode 4 are similar, which are slightly narrower than that derived from Mode 1, especially in low flow conditions. This is consistent with the finding from parameter identification results under different calibration modes that parameter set optimized from single-site calibration (Mode 1) were found to have the largest uncertainty ranges while parameter sets estimated from multi-site calibration (Mode 2- Mode 4) have similarly narrow uncertainty ranges. This indicates that including discharge observations from internal site(s) into predictive analysis can decrease prediction uncertainty and the extent depends on the information content of the added data. There are no significant differences in predictive uncertainty ranges derived from all four predictive analysis modes (Mode 1- Mode 4). Thus, in this nested catchment where each part of the catchment is hydrologically correlated, including discharge observations from internal site(s) into predictive analysis rather than only using discharge measurements from catchment outlet does not have substantial influence on predictive uncertainty. Based on the small prediction uncertainty ranges estimated from all four different predictive analysis modes, it implies that the parameter non-uniqueness does not result in high level of

predictive uncertainty. Therefore, the HYPE model structure is physically sound and robust in terms of simulated hydrological processes and parameter set.

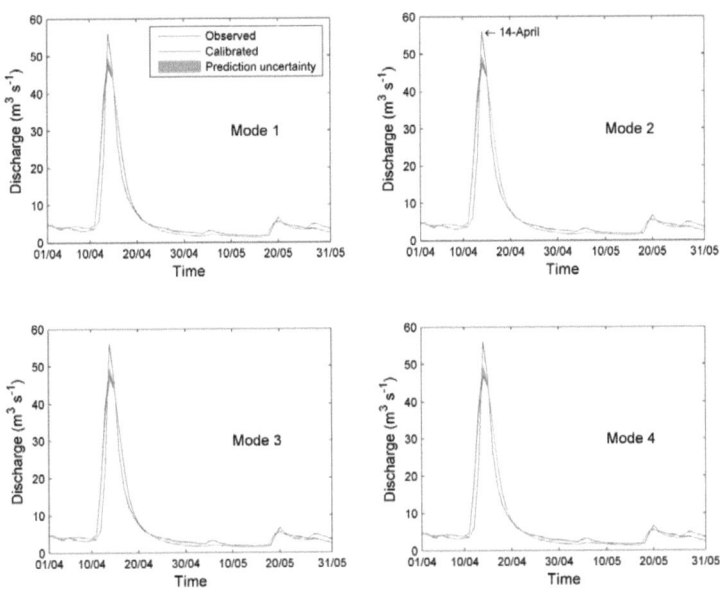

Figure 5.3 Observed and best simulated discharge as well as predictive uncertainty range at the catchment outlet (gauging station Hausneindorf) for a two-month period (April-May of 1994) estimated from four different calibration and predictive analysis modes (Mode 1- Mode 4).

Through comparing results of parameters identification, model performances and discharge prediction uncertainty at catchment outlet derived different calibration and predictive analysis modes that involve discharge observations from different number of gauging stations.We noticed that (i) Posterior parameters' uncertainties decrease when discharge observations from internal sites were included for parameter calibration (i.e. multi-site calibration); while the extent of this effect depends on the information content of the added data. (ii) There is no obvious improvement of model performance when multi-site calibration was implemented compared with single-site calibration because of the "nested" nature of the Selke catchment and high correlation between observed discharges from different parts of the catchment; small trade-offs between three gauged sub-basinsoccurred when multi-site calibration was applied rather than single-site calibration, which is reflected by different behaviors of water balance (PBIAS). (iii) The parameter sets estimated from predictive analysis are statistically similar to the corresponding calibration optimized parameter set, which indicates that the model structure of HYPE is robust and parameters are highly identifiable. (iv) The HYPE model has similar performances using estimated parameter sets from predictive analysis as with calibrated parameter sets, which implies that the model structure of HYPE is physically sound and robust. (v) The discharge prediction uncertainties estimated from both single-site predictive analysis and multi-site predictive analysis are small, which indicates that parameter non-uniqueness does not result in high level of predictive uncertainty. (vi) The discharge prediction uncertainty decreases only slightly when discharge measurements from either internal site or both sites were included for predictive analysis, which is attributed to the high correlation between different parts of the catchment in terms of hydrological processes.

5.1.1.2 Calibration and predictive analysis using DREAM$_{(ZS)}$

DREAM$_{(ZS)}$ was combined with the HYPE model to implement hydrological parameter optimization and uncertainty analysis at Selke catchment. Single-site calibration (i.e. using discharge observations only from catchment outlet (gauging station Hausneindorf)) and multi-site calibration (i.e. using discharge observations from all three monitoring sites (gauging stations Hausneindorf, Silberhuette, and Meisdorf)) were implemented and compared.

The theory of using DREAM$_{(ZS)}$ for parameter calibration and uncertainty analysis has been described in detail in Chapter 2 (Section 2.4.2) and Appendix 3. The HYPE model was set up for discharge simulation following the same procedures applied in section 5.1.1.1. In brief, the model was calibrated for the period 1994-1999 and one year (1993-1994) was run as warming up and excluded from model evaluation. The same hydrological parameters listed in Table 4.3 were chosen for calibration using DREAM$_{(ZS)}$. Initial ranges for all parameters were defined as those specified in PEST run (Table 4.3), except that the initial range of rivvel was defined as 0.1-1. The reason is that DREAM$_{(ZS)}$ was found to stop after some runs when larger initial range of rivvel was defined, such as 0.001-1.0 or 0.01-1.0. This may be caused by the HYPE model crash in a global search process when the parameter *rivvel* obtain an unrealistic value. Standard least squares (SLS) were chosen as the likelihood function. Three Markov chains ($N=3$) were specified for the global search; the total number of function evaluations was defined as 10000. All other parameters for DREAM$_{(ZS)}$ run were defined referring to Laloy and Vrugt (2012).

Parameter identification

Parameter calibration results from single-site calibration (SSC) and multi-site calibration (MSC), including optimized values (OPV), standard deviations (STD) and mean values of each calibrated parameter are given in Table 5.4. The convergence of calibrated parameters derived from SSC and MSC are shown in Figure 5.4. Taking the most sensitive hydrological parameter *cevp2* (potential evapotranspiration coefficient of agricultural land) as an example, chain convergence of SSC and MSC are presented in Figure 5.5. Histograms of marginal distributions of individual parameters (pdf) derived from SSC and MSC are shown in Figure 5.6.

From the convergence of calibrated parameters (Figure 5.4), it is noted that with MSC, sampled chains have better and faster convergence than SSC. Parameter cevp2 show more constant evolution and constrained sampling space derived from MSC comparing with SSC (Figure 5.5).This indicates that multi-site calibration increases constraints on hydrological processes and therefore improves evolution of parameter posterior distribution. Similar patterns were found for the other four parameters. From the histograms of marginal distributions of individual parameters (Figure 5.6), the shape of histogram for each parameter derived from SSC and MSC are different from each other, reflecting the uncertainty of parameter posterior distributions estimated from $DREAM_{(ZS)}$. However, the parameters estimated from MSC have much smaller uncertainties indicated by lower standard deviations (Table 5.4) and narrower posterior distributions (Figure 5.6) compared with SSC. Therefore, multi-site calibration improves parameter identification since spatial variability of hydrological processes related to change of land use and soil is accounted when discharge observations from internal sites are added for calibration. This finding is consistent with that derived from comparison of SSC and MSC using PEST (Figure 5.1). Optimized parameter values from SSC and MSC are similar, indicating that the structure of the HYPE model for hydrological simulation is robust and the relevant parameters are highly identifiable.

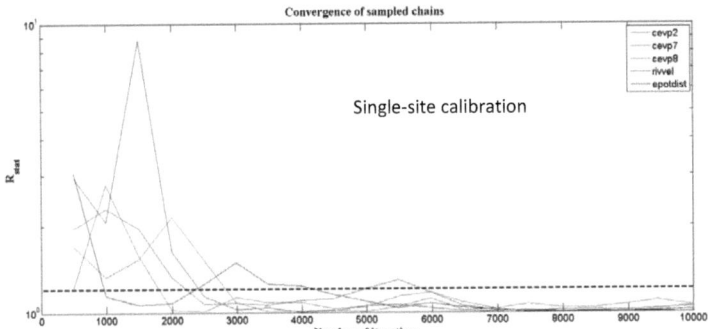

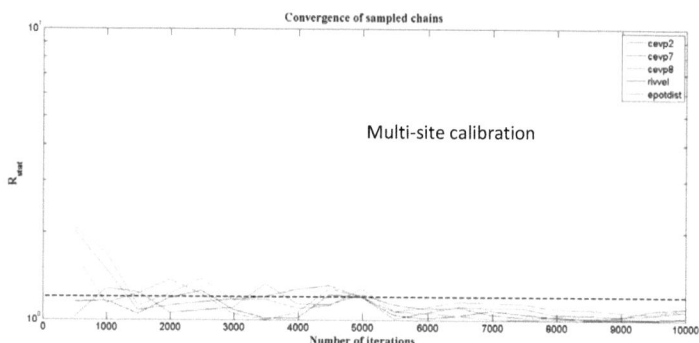

Figure 5.4 Convergence of calibrated parameters derived from single-site calibration (top panel) and multi-site calibration (bottom panel) (the dashed line represents the threshold ($\hat{R}_j < 1.2$) that declares convergence of a parameter to a limiting distribution (Laloy and Vrugt, 2012)).

Table 5.4 Parameter calibration results from single-site calibration (SSC) and multi-site calibration (MSC).

Parameter	Single-site calibration (SSC)			Multi-site calibration (MSC)		
	Optimized value	Mean value	Standard Dev	Optimized value	Mean value	Standard Dev
cevp2	0.319	0.326	0.012	0.308	0.316	0.007
cevp7	0.164	0.157	0.018	0.169	0.170	0.002
cevp8	0.123	0.136	0.016	0.123	0.125	0.003
rivvel	0.216	0.216	0.003	0.205	0.206	0.002
epotdist	8.788	9.149	0.359	8.992	9.425	0.423

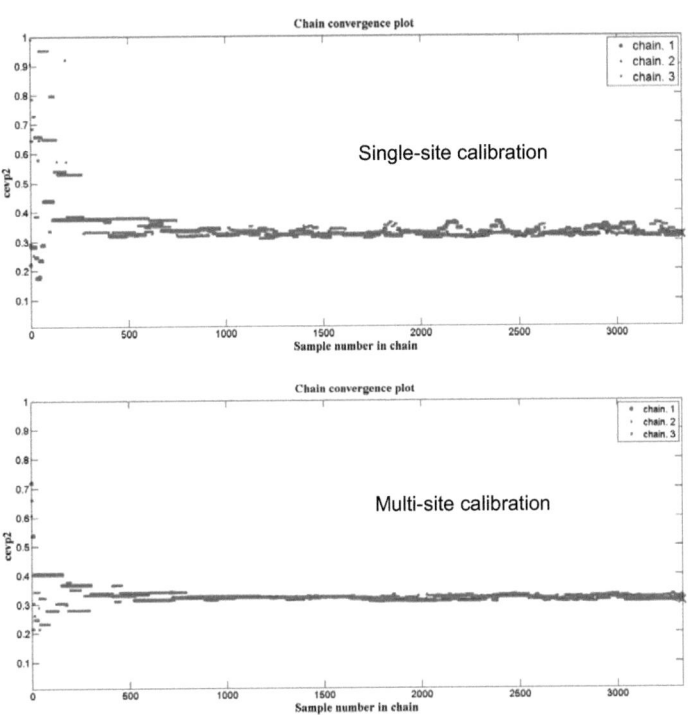

Figure 5.5 Chain convergence of parameter *cevp2* (potential evapotranspiration coefficient of agricultural land) derived from single-site calibration (top panel) and multi-site calibration (bottom panel).

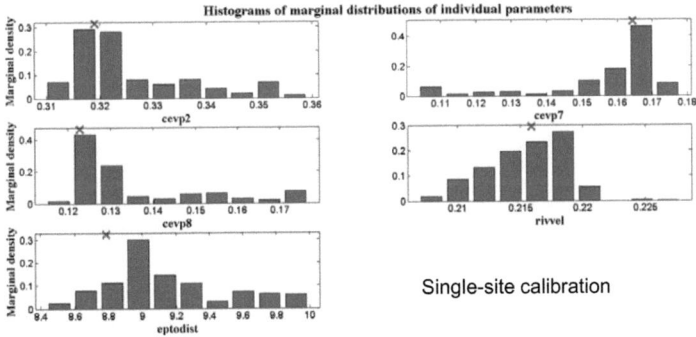

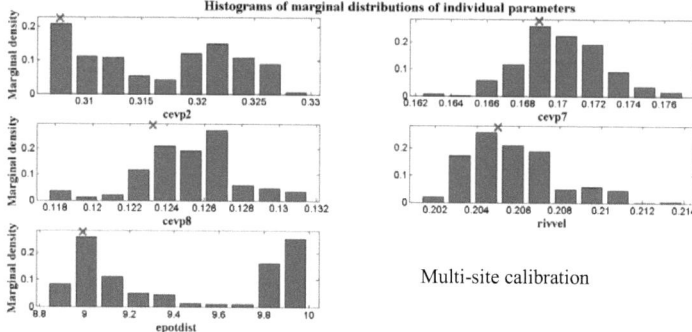

Multi-site calibration

Figure 5.6 Histograms of marginal distributions of individual parameters (pdf) derived from single-site calibration (top panel) and multi-site calibration (bottom panel).

Model performance

Table 5.5 shows the statistical model performance on discharge simulations for calibration and validation modes using the parameter sets calibrated from single-site calibration (SSC) and multi-site calibration (MSC). The HYPE model has similar performance at each gauging station using parameter sets optimized from SSC and MSC in both calibration and validation periods. The only difference is noticed in model performance of water balance (PBIAS) comparing MSC and SSC attributed to some trade-offs between sub-basins when multi-site calibration is applied. This was also found in comparison of model performance derived from SSC and MSC using PEST (Figure 5.2). Therefore, multi-site calibration does not improve model performance on discharge simulations substantially in this nested catchment comparing with single-site calibration, in which hydrological processes and observed discharges at different parts are highly correlated.

Table 5.5 Statistical model performance on discharge simulations during calibration and validation periods using the parameter sets optimized from single-site calibration (SSC) and multi-site calibration (MSC) (MAE and RMSE stand for mean absolute error and root mean squared error, respectively)

	Calibration period (1994-1999)				Validation period (1999-2004)			
	NSE	PBIAS (%)	MAE ($m^3 s^{-1}$)	RMSE ($m^3 s^{-1}$)	NSE	PBIAS (%)	MAE ($m^3 s^{-1}$)	RMSE ($m^3 s^{-1}$)
	Single-site calibration (SSC)							
Silberhuette	0.879	-6.21	0.329	0.792	0.901	-11.08	0.264	0.498
Meisdorf	0.877	-4.55	0.373	0.859	0.901	-0.49	0.269	0.555
Hausneindorf	0.863	1.41	0.513	1.129	0.873	14.65	0.456	0.806

				Multi-site calibration (MSC)				
Silberhuette	0.883	-5.91	0.324	0.780	0.904	-10.78	0.261	0.241
Meisdorf	0.878	-4.38	0.373	0.855	0.900	-0.309	0.270	0.557
Hausneindorf	0.862	2.29	0.519	1.133	0.868	15.48	0.466	0.822

Predictive uncertainty

Figure 5.7 showed the discharge prediction uncertainty at the catchment outlet (discharge gauging station Hausneindorf) estimated from single-site calibration (SSC) and multi-site calibration (MSC) using DREAM$_{(ZS)}$ for the period April-May of 1994 . The parameter uncertainty (black ranges) stands for the prediction uncertainty caused by parameter non-uniqueness related to model complexity and parameter interaction. The total uncertainty represents prediction uncertainty attributed to various sources (e.g. model structure uncertainty, parameter uncertainty, input data uncertainty and uncertainty of the observed responses).

Parameter uncertainty is small, while total uncertainty is much larger; observed discharges lie close to the narrow parameter uncertainty ranges except for the extraordinary peak flow event of 14.04.1994 (Figure 5.7). Uncertainty ranges (parameter uncertainty and total uncertainty) estimated from MSC are relatively narrower than those derived from SSC. Statistically, the observed discharges for this two-month period have the range of 1.73-56.03 $m^3 s^{-1}$; the parameter uncertainty ranges estimated from SSC and MSC are 1.34-48.21 $m^3 s^{-1}$ and 1.36-48.08 $m^3 s^{-1}$; the total uncertainty estimated from SSC and MSC are -0.92-49.30 $m^3 s^{-1}$ and -0.48-49.14 $m^3 s^{-1}$. The lower prediction uncertainty ranges estimated from MSC than SSC indicate that increasing discharge observations from internal sites of thecatchment decreases predictive uncertainty and thus increase prediction accuracy. The extent of these effects depends on the additional information contents of observations from internal sitesrelated to catchment characteristics. The narrow parameter uncertainty estimated from SSC and MOC using DREAM$_{(ZS)}$ are consistent with those estimated from predictive analysis using PEST (Figures 5.7 and 5.3).

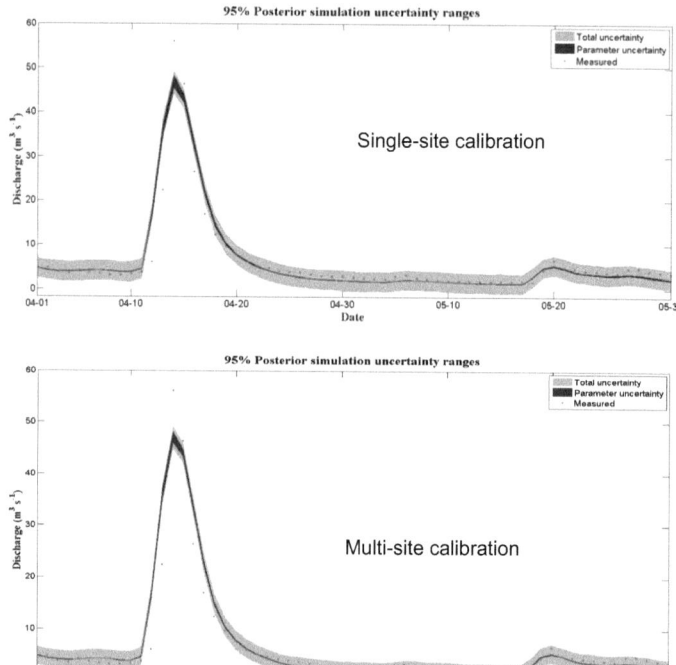

Figure 5.7 Discharge prediction uncertainty (parameter uncertainty (black area) and total uncertainty (grey area)) at the catchment outlet (discharge gauging station Hausneindorf) for the period April-May of 1994 estimated from single-site calibration (top panel) and multi-site calibration (bottom panel) using DREAM$_{(ZS)}$

5.1.2 IN concentration simulation

5.1.2.1 Calibration and predictive analysis using PEST

<u>Parameter identification</u>

Following the similar calibration and predictive analysis procedures for runoff simulation applied in section 5.1.1.1, PEST was combined with the HYPE model to implement single-site calibration and predictive analysis (SSCP) which uses IN concentration measurements only from catchment outlet (gauging station Hausneindorf) and multi-site calibration and predictive analysis (MSCP) that uses IN concentration measurements from all three sampled sites (gauging stations Hausneindorf, Silberhuette, and Meisdorf) for IN concentration simulation. Model calibration results derived from both procedures, such asparameter identification, model performance and predictive uncertainty were compared. Model calibration and predictive analysis were run for the period 1994-1999 and one year (1993-1994) was used as model warming up and excluded for model performance evaluation. The temporal and spatial model performances on simulations of IN concentration and daily IN loads using parameter sets optimized from MSCand SSC were compared. In addition, model performance using parameter sets estimated from predictive analysis (minimum and maximum predictive analysis) were compared with that derived using the corresponding calibrated parameter set.The same nitrogen-process parameters, initial values and ranges as those listed in Table 4.3 were chosen for calibration.

The parameter calibration and predictive analysis results including relative composite sensitivity (RCS), optimized parameter values, posterior 95% confidence limits, estimated parameter values and standard deviation derived from SSCP and MSCPare given in Table 5.6. The three most sensitive parameters determined from SSC and MSC are same.They are *uptsoil102*, *uptsoil108* (Nutrient uptake in the uppermost soil layer for agricultural land and mixed forest) and *denitr* (Denitrification rate in soil). There is no significant difference between the parameter values optimized from SSC and MOC. This indicates that the water quality parameters of the HYPE model are highly identifiable.In the multi-site calibration, the contributions to the objective function(sum of squared weighted residuals) from the IN concentration observation groups "Hausneindorf", "Silberhuette" and "Meisdorf" are 182.8, 14.66 and 18.81, respectively. The high similarity of parameter sets estimated from SSC and MSC is also attributed to the dominant role of observation group "Hausneindorf" in objective function related to higher number of observations and different patterns of IN concentration dynamics comparing with observation groups "Silberhuette" and "Meisdorf". Considering parameter posterior uncertainty, 95% confidence limits of all parameters estimated from MSC are narrower than those estimated from SSC except for parameter uptsoil107 (Table 5.6). This indicates that multi-site calibration enables to improve water quality parameter identification by decreasing posterior uncertainty ranges. Through comparing the parameter sets estimated from minimum and maximum predictive analysis with the corresponding optimized parameter set under both SSCP and MSCP using t-test, they are statistically the same because there is no significant difference between the estimated parameter values and 95% confidence limits overlap each other. Therefore, the optimized parameters can be considered as global optimal.

Table 5.6 Calibration and predictive analysis results of nitrogen-process parameters derived from single-site calibration and predictive analysis (SSCP) and multi-site calibration and predictive analysis (MSCP) (RCS- relative composite sensitivity).

Parameter	RCS	Calibration mode		Minimum prediction		Maximum prediction	
		Optimized value	95% confidence limits	Estimated value	Standard deviation	Estimated value	Standard deviation
Single-site calibration and predictive analysis (SSCP)							
denitr	0.106	0.023	0.012-0.035	0.032	0.008	0.017	0.005
denitw	0.003	0.0001	-0.002-0.002	0.0001	0.0009	0.0005	0.0009
wprod	0.006	0.003	-0.007-0.013	0.0001	0.006	0.005	0.006
uptsoil102	0.46	1.0	0.907-1.093	1.0	0.054	1.0	0.045
uptsoil107	0.012	1.0	-2.222-4.222	0.990	3.026	1.0	0.622
uptsoil108	0.278	1.0	0.839-1.161	1.0	0.184	1.0	0.089
Multi-site calibration and predictive analysis (MSCP)							
denitr	0.064	0.022	0.017-0.028	0.030	0.004	0.017	0.002
denitw	0.002	0.0001	-0.001-0.001	0.0001	0.0006	0.0006	0.0006
wprod	0.009	0.006	-0.0009-0.013	0.003	0.004	0.007	0.004
uptsoil102	0.267	1.0	0.944-1.056	1.0	0.032	1.0	0.027
uptsoil107	0.0002	0.897	-47.75-49.54	0.897	50.80	0.897	12.977
uptsoil108	0.137	0.969	0.867-1.071	0.994	0.039	0.975	0.053

Model performance

Statistical model performances on simulations of IN concentration and daily IN loadsat all three gauging stations using parameter sets optimized from SSC and MSC in calibration (1994-1999) and validation (1999-2004) periods are given in Table 5.7. Considering IN concentration simulations, model performances at the gauging stations Silberhuette and Meisdorf derived using parameter set calibrated from MSC are better than those obtained from SSC for both calibration and validation modes in terms of R^2 and MAE. Model performances at the catchment outlet (gauging station Hausneindorf) obtained under both calibration procedures are similar. This indicates that adding the IN concentration measurements from internal sites improves spatial prediction accuracy because the spatial variability of nitrogen transport and transformation related to catchment heterogeneity can be accounted for with increasing spatially distributed information and constraints.

Considering daily IN loads simulations, model performances derived from SSC and MSC are similar for both calibration and validation modes at each gauging station in terms of R^2, NSE and RSR. The HYPE model had slightly different performance obtained from SSC and MSC at each gauging station in calibration and validation periods in terms of nitrogen load balance reflected by PBIAS (Table 5.7). This is attributed to trade-off between different observation groups during calibration process using MSC. Overall better model performances during validation period derived from MSC are noted comparing with those obtained from SSC in terms of nitrogen loads balance. The better and more robust model performances on simulations of IN concentration and loads indicate that multi-site calibration is more reasonable to implement proper spatial distributed predictions.

Table 5.7 Statistical model performances on simulations of IN concentration and daily IN loads using parameter sets optimized from single-site (SSC) and multi-site (MSC) calibration using PEST.

Variable	Criterion	Calibration						Validation					
		Silberhuette		Meisdorf		Hausneindorf		Silberhuette		Meisdorf		Hausneindorf	
		SSC	MSC	SSC	MSC	SSC	MSC	SSC	MSC	SSC	MSC	SSC	MSC
IN concentration	R^2	0.70	0.74	0.57	0.63	0.08	0.08	0.39	0.43	0.28	0.33	0.06	0.05
	$MAE\ (mg\ L^{-1})$	0.504	0.450	0.520	0.475	0.873	0.874	0.764	0.685	0.548	0.491	0.967	0.980
Daily IN load	R^2	0.92	0.92	0.83	0.83	0.73	0.72	0.83	0.83	0.91	0.91	0.84	0.84
	NSE	0.89	0.89	0.82	0.82	0.71	0.69	0.83	0.83	0.88	0.88	0.50	0.50
	PBIAS	-8.41	-9.83	-11.29	-13.77	6.38	5.93	3.72	1.88	11.63	8.53	40.57	40.08
	RSR	0.33	0.33	0.42	0.43	0.54	0.55	0.41	0.41	0.35	0.34	0.71	0.71

Predictive uncertainty

Figure 5.8 shows IN concentration predictive uncertainty induced byparameter non-uniqueness at catchment outlet (gauging station Hausneindorf) for a two-month period (February and March of 1996) estimated from single-site (SSCP) and multi-site (MSCP) calibration and predictive analysis. Based on the theory of predictive analysis using PEST, the key prediction was chosen as the peak value of IN concentration measured at the gauging station Hausneindorf on 08.01.1996 with the observed value of 5.27 mg L^{-1}. The calculated IN concentrations using the best optimized parameter set are located inside the uncertainty ranges shaped by minimum- and maximum predictions under both MSCP and SSCP.The prediction uncertainty ranges derived from both calibration and predictive analysis procedures are small. This indicates that parameter non-uniqueness does not result in high level of prediction uncertainty. It is important to note that the uncertainty ranges of IN concentrations estimated from MSCP is much narrower than that estimated from SSCP (Figure 5.8). This verifies that including IN concentration measurements from internal sites for calibration and predictive analysis can decrease prediction uncertainty. This is because nitrogen-process parameters are better identified associating with smaller uncertainty under MSCP comparing with SSCP in this heterogeneous catchment (Table 5.6).

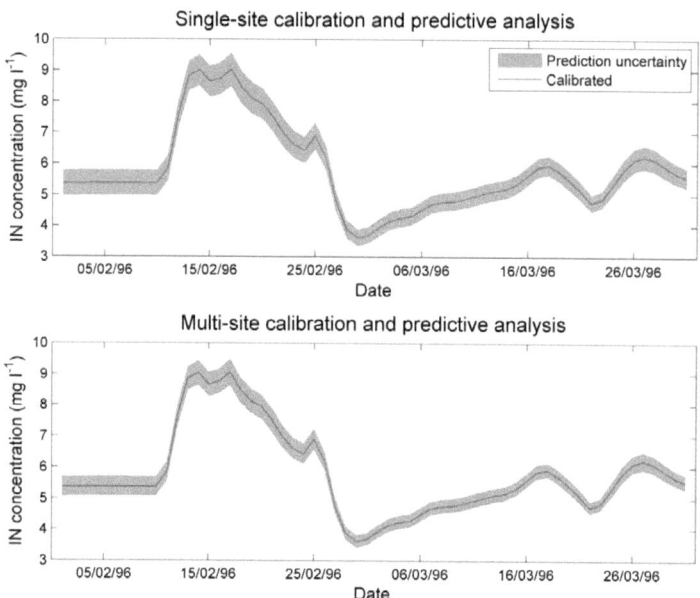

Figure 5.8 Prediction uncertainty of IN concentrations at the catchment outlet (gauging station Hausneindorf) for a two-month period (February and March of 1996) estimated from single-site (top panel) and multi-site (bottom panel) calibration and predictive analysis

5.1.2.2 Calibration and predictive analysis using DREAM$_{(ZS)}$

<u>Parameter identification</u>

In this part DREAM$_{(ZS)}$ was combined with the HYPE model to implement single-site and multi-site calibration and uncertainty analysis on nitrogen-process parameters. Nitrogen-process parameters and initial ranges listed in Table 4.3 were specified for calibration. The number of model runs was defined as 10000. Three Markov chains (N=3) were specified for the global search. Standard least squares (SLS) were chosen as likelihood function. The histograms of marginal distributions of five nitrogen-process parameters derived from single-site and multi-site calibration are presented in Figure 5.9. The parameter calibration results, including optimized values, mean values and standard deviations are given in Table 5.8. From Figure 5.9 and Table 5.8, it is noted that parameters' uncertainty ranges estimated from multi-site calibration are smaller than those derived from single-site calibration, except for the parameter *uptsoil107*. The exception of *uptsolil107* is probably attributed to its low sensitivity according to Table 5.6. There is no significant difference between the values of parameters *uptsoil102*, *uptsoil108* and *denitr* estimated from SSC and MSC. However, for the insensitive parameter *uptsoil107*, the optimized values from SSC and MSC are different. The lower identifiability of the parameter *uptsoil107* is also reflected by its larger standard deviation (Table 5.8). The optimized value of uptsoil107 estimated from multi-site calibration is considered to be reasonable as it is similar to that calibrated under both SSC and MSC using PEST (Table 5.6).

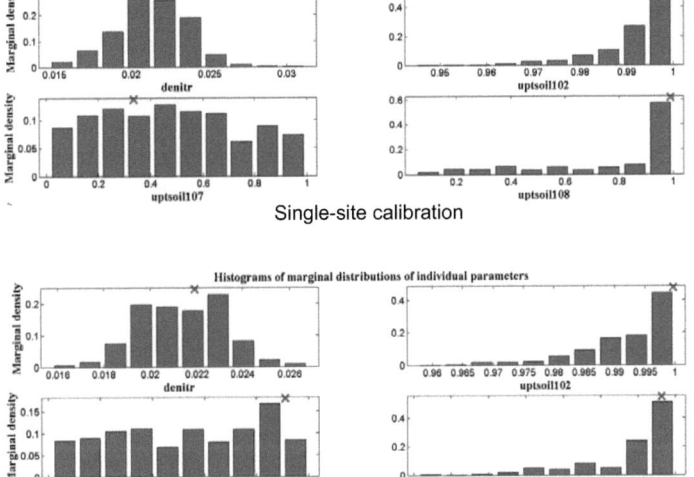

Single-site calibration

Multi-site calibration

Figure 5.9 Histograms of marginal distributions of four nitrogen-process parameters derived from single-site calibration (top panel) and multi-site calibration (bottom panel).

Table 5.8 Nitrogen-process parameter calibration results of single-site calibration (SSC) and multi-site calibration (MSC).

Parameter	Single-site calibration (SSC)			Multi-site calibration (MSC)		
	Optimized value	Mean value	Standard Dev	Optimized value	Mean value	Standard Dev
denitr	0.022	0.021	0.0023	0.022	0.021	0.0018
uptsoil102	1.0	0.99	0.0081	1.0	0.99	0.0075
uptsoil107	0.34	0.47	0.269	0.91	0.53	0.288
uptsoil108	0.99	0.79	0.268	0.96	0.88	0.128

Model performance

The statistical model performances on simulations of IN concentration and daily IN loads with parameter sets estimated from single-site and multi-site calibration using DREAM$_{(ZS)}$ at all three gauging stations during calibration (1994-1999) and validation (1999-2004) periods are given in Table 5.9. Similar as the findings derived from comparison of model performances under SSC and MSC using PEST (Table 5.7), the HYPE model had better performances on IN concentration simulations using parameters optimized from MSC than those derived from SSC at internal sites Silberhuette and Meisdorf for both calibration and validation modes. Model performances on daily IN loads simulations at gauging stations Silberhuette and Meisdorf obtained from MSC are slightly better than those derived from SSC during both calibration and validation periods in terms of nitrogen load balance. These findings verify again the importance of IN concentration measurements from internal sites for improving model parameter identification and increasingspatial prediction accuracy. Improvements of MSC over SSC are for IN concentrations much higher tithan for hydrological simulations. This may be attributed to the fact that (i) the number of observations for IN concentrations are much smaller than for discharge measurements; (ii) the correlation between IN concentration measurements derived from different gauging stations are much lower than that for discharge observations obtained from different gauging stations.

Table 5.9 Statistical model performance on simulations of IN concentration and daily IN loads at all three gauging stations during calibration (1994-1999) and validation (1999-2004) periods using the parameter sets optimized from single-site and multi-site calibration using DREAM$_{(ZS)}$.

Variable	Criterion	Calibration						Validation					
		Silberhuette		Meisdorf		Hausneindorf		Silberhuette		Meisdorf		Hausneindorf	
		SSC	MSC	SSC	MSC	SSC	MSC	SSC	MSC	SSC	MSC	SSC	MSC
IN concentration	R^2	0.68	0.74	0.55	0.62	0.08	0.08	0.34	0.43	0.25	0.33	0.06	0.05
	$MAE\ (mg\ L^{-1})$	0.523	0.451	0.522	0.475	0.874	0.869	0.753	0.695	0.545	0.502	0.985	0.990
	R^2	0.91	0.92	0.83	0.83	0.72	0.72	0.81	0.83	0.91	0.91	0.84	0.84
Daily IN load	NSE	0.87	0.89	0.81	0.81	0.70	0.69	0.80	0.83	0.88	0.88	0.49	0.48
	PBIAS	-10.79	-9.20	-13.84	-13.13	6.20	6.41	2.95	2.86	10.03	9.68	41.65	41.65
	RSR	0.36	0.33	0.43	0.43	0.55	0.55	0.44	0.41	0.34	0.35	0.71	0.72

Predictive uncertainty

The prediction uncertainty (parameter uncertainty and total uncertainty) of IN concentrationsat catchment outlet estimated from single-site and multi-site calibration and uncertainty analysis were shown in Figure 5.10. Under single-site calibration and uncertainty analysis (SSCP), a total number of 129 streamwater IN concentration measurements (red dots in Figure 5.10) from gauging station Hausneindorf during 1994-1999 (biweekly to weekly interval) were considered. While for multi-site calibration and uncertainty analysis (MSCP), a total number of 229 streamwater IN concentration measurements (50 from Silberhuette, 50 from Meisdorf and 129 from Hausneindorf) were used for the same period (1994-1999). The total uncertainty ranges are relatively large under both SSCP and MSCP. Most IN concentration observations lie inside the total uncertainty range under both SSCP and MSCP. Both total uncertainty and parameter uncertainty derived from MSCP are narrower than those estimated from SSCP. This indicates that adding IN concentration measurements from internal sites into calibration and uncertainty analysis enables to decrease prediction uncertainty attributed to increased constraints on water quality processes.

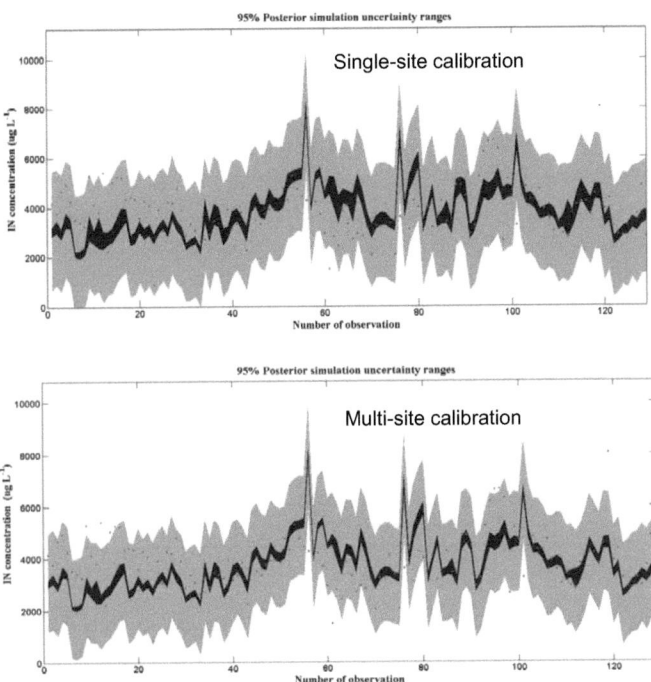

Figure 5.10 Prediction uncertainty of IN concentrations (parameter uncertainty (black ranges) and total uncertainty (grey ranges)) as well as observed IN concentrations (red dots) at the catchment outlet (gauging station Hausneindorf) estimated from single-site (top panel) and multi-site (bottom panel) calibration and uncertainty analysis for the calibration period (1994-1999) using DREAM$_{(ZS)}$.

5.2 Temporal resolution effect of calibration data

The effects of temporal resolution of nitrate-N concentration observations on model parameter identification, model performance, accuracy of nitrate-N loads prediction, and prediction uncertainty are investigated. The HYPE model was set up for IN simulation on a two-year period using split-sample test (i.e. 2011 for calibration and 2012 for validation). PEST and DREAM$_{(ZS)}$ were combined with the HYPE model to implement parameter automatic calibration and predictive analysis. Bi-Weekly nitrate-N concentration Samplings (BWSC) and Daily Averages of 15-min step online nitrate-N concentration measurements (DAOC) were compared in terms of parameter identification, model performance and prediction uncertainty.

5.2.1 Discharge simulation

In discharge simulation for the period 2011-2012, the hydrological parameter values were directly obtained from multi-site calibration during the period 1994-1999 using PEST (Mode 4 of Table 5.1). The purpose is to test the temporal transferability of the model structure and parameter set between different climatic and hydrological conditions. The simulated and observed discharges at three gauging stations (Silberhuette, Meisdorf and Hausneindorf) for the period 01.11.2010-31.12.2012 are shown in Figure 5.11. The statistical model performance on discharge simulations of 2011 and 2012 are given in Table 5.10.

Table 5.10 Model performances on discharge simulations at three gauging stations (Silberhuette, Meisdorf and Hausneindorf) for 2011 and 2012 using the hydrological parameter set optimized from multi-site calibration during the period 1994-1999 using PEST.

Sub-basin	2010-2011			2012		
	NSE	Annual runoff (mm/y)		NSE	Annual runoff (mm/y)	
		Simulated	Observed		Simulated	Observed
Silberhuette	0.94	264.4	271.0	0.80	220.9	178.3
Meisdorf	0.83	212.3	262.6	0.78	169.3	143.6
Hausneindorf	0.88	140.7	136.6	0.72	93.6	82.2

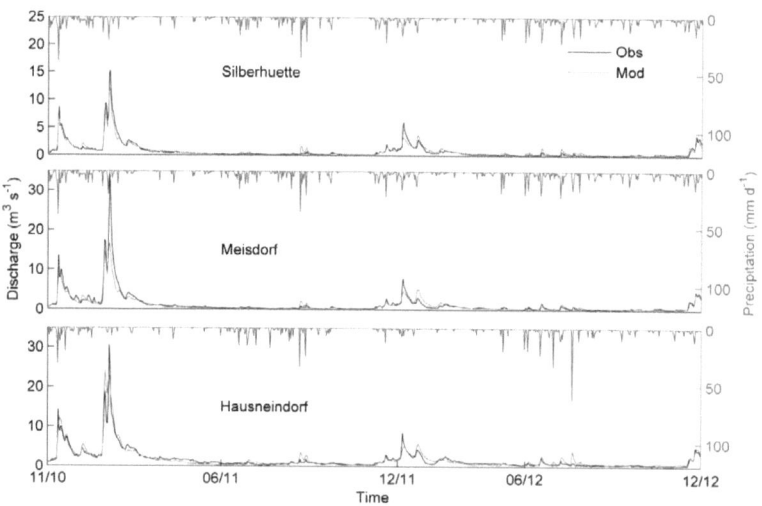

110

Figure 5.11 Discharge simulation results at three gauging stations (Silberhuette, Meisdorf and Hausneindorf) for the period 01.11.2010-31.12.2012 using the hydrological parameter set optimized from multi-site calibration during the period 1994-1999 using PEST.

The observed hydrographs at all three gauging stations for the whole simulation period (01.11.2010-31.12.2012) are represented well using the parameter set calibrated during the historical period (1994-1999). This indicates that the well identified hydrological parameters of HYPE are transferable temporally between various climatic and hydrological conditions. Most peak flows and low flows are captured by the model, except for the extraordinary high flow events occurred in January of 2011, which is noted by substantial underestimation of peak discharges at the gauging station Meisdorf. The considerable decrease of observed catchment specific annual runoff from 2011 to 2012 is noted from Table 5.10, due to the extraordinary flood event occurred in January of 2011. The discrepancies between the simulated and observed annual runoff are mainly attributed to the input uncertainty (precipitation) because precipitation monitored from four rainfall stations may not properly represent the spatial variability of precipitation. On the whole, the HYPE model simulated discharge satisfactorily for the whole period (2010-2012) and at all three gauging stations.

5.2.2 Calibration and predictive analysis using PEST

Parameter identification

For water quality parameter optimization the same six nitrogen-process parameters, initial values and ranges listed in Table 4.3 were chosen for calibration and predictive analysis against biweekly nitrate-N concentration measurements (BWSC) and daily averages of nitrate-N observations (DAOC) using PEST. Multi-site calibration (i.e. using nitrate-N concentration measurements from all three gauging stations) was implemented under both BWSC and DAOC for the period 01.11.2010-31.12.2011. The model was validated for the period 01.01.2012-31.12.2012. Model identification results of optimized/estimated values and posterior uncertainty ranges derived from calibration and predictive analysis under BWSC and DAOC were compared.

The parameter identification results derived from calibration and predictive analysis based on BWSC and DAOC using PEST are given in Table 5.11. The ranks of parameter sensitivity obtained from calibration under BWSC and DAOC are nearly identical. The most sensitive parameter is *uptsoil102* (Fraction of nutrient uptake in the uppermost soil layer), which is consistent with the finding from former parameter sensitivity analysis during the period 1994-1999. This indicates the dominant role of plant uptake in nitrogen balance. The second most sensitive parameter is *denitw* (Denitrification rate in water), which is different from former parameter sensitivity analysis results. This is because parameter sensitivity may change depending on variations of climate (e.g. precipitation and temperature) and hydrological conditions (Lord et al., 2002; Van Griensven and Bauwens, 2003).

Except for the parameter *denitw*, there are no significant differences between the parameter values optimized from calibrations under BWSC and DAOC and their 95% confidence limits overlay each other (Table 5.11). Parameter *denitw* has higher estimated value from calibration under DAOC than that optimized from calibration under BWSC. This may be explained by the fact that much more low-flow nitrate-N concentration observations were included in calibration under DAOC when in-stream denitrification is important comparing with calibration under BWSC. Therefore, water quality parameter calibration using bi-weekly nitrate-N concentration measurements has the risk of underestimating in-stream retention effects (e.g. denitrification). Parameter posterior uncertainty ranges obtained from calibration under DAOC are narrower than those estimated from calibration under BWSC. This indicates that water quality parameter calibration using high temporal resolution (daily) of nitrate-N concentration measurements can improve parameter identification by decreasing parameter posterior uncertainty. Comparing with regular biweekly nitrate-N concentration measurements, continuous daily nitrate-N observations have the advantage of capturing detailed variations of weather, hydrological processes and nitrate-N concentrations, which helps for identifying nitrogen-process parameters with high accuracy. Parameter set estimated from maximum predictive analysis is statistically the same as the corresponding optimized parameter set under both BWSC and DAOC. In minimum predictive analysis, only the parameter prod estimated under BWSC and *uptsoil108* estimated under DAOC is statistically different from the respective optimized value. This verifies the high identifiability of water quality parameters of the HYPE model.

Table 5.11 Nitrogen-process parameter identification results of calibration and predictive analysis against bi-weekly nitrate-N concentration measurements (BWSC) and daily averages of nitrate-N concentration observations (DAOC) (RCS- relative composite sensitivity).

Parameter	RCS	Calibration mode		Minimum prediction		Maximum prediction	
		Optimized value	95% confidence limits	Estimated value	Standard deviation	Estimated value	Standard deviation
Bi-weekly nitrate-N concentration measurements (BWSC)							
denitr	0.015	0.001	-0.003-0.005	0.001	0.0022	0.001	0.0021
denitw	0.124	0.0002	0.0001-0.0003	0.0003	0.00006	0.0002	0.00005
wprod	0.0003	0.0001	-0.01-0.01	0.0036	0.0055	0.0001	0.005
uptsoil102	0.779	1.0	0.944-1.056	1.0	0.032	1.0	0.029
uptsoil107	0.073	1.0	0.475-1.525	1.0	0.30	0.99	0.26
uptsoil108	0.020	0.597	-0.137-1.330	0.47	0.130	0.72	0.30
Daily averages of nitrate-N concentration observations (DAOC)							
denitr	0.008	0.002	0.0005-0.004	0.002	0.0008	0.0028	0.0008
denitw	0.039	0.004	0.0038-0.0049	0.0048	0.0003	0.0036	0.0003
wprod	0.00006	0.0001	-0.005-0.005	0.0001	0.0022	0.0001	0.0023

uptsoil102	0.232	1.0	0.982-1.018	1.0	0.0088	1.0	0.0088
uptsoil107	0.0269	1.0	0.814-1.186	1.0	0.066	0.99	0.065
uptsoil108	0.0267	0.70	0.590-0.811	0.40	0.12	0.64	0.086

Model performance

The observed and simulated IN concentration and daily IN loads under calibration (2010-2011) and validation (2012) against biweekly nitrate-N concentration measurements (BWSC) and daily averages of nitrate-N concentration observations (DAOC) are presented in Figures5.12-5.15. The corresponding statistical model performances on simulations of IN concentration and daily IN loads in calibration and validation periodsunder BWSC and DAOC are given in Table 5.12. The seasonal dynamics of IN concentrations are well captured in the upper-stream sites (gauging stations Silberhuette and Meisdorf) under both BWSC and DAOC. IN concentrations show proportional relationship with discharge characterized by high concentrations in peak flow events in winter and low concentrations during low flow conditions in summer. As explained in IN simulations for historical period (1994-1999) in Chapter 3 (Section 3.1.3.1), these are the combined effects of hydrological and biogeochemical processes.

Due to underestimation of peak flows in winter (2010.11-2011.01), IN concentrations are underestimated, especially in the gauging station Meisdorf. This indicates the importance of correct prediction of peak flows for IN concentrations simulation. In the similar mesoscale catchment (Weida), Shrestha *et al.* (2013) pointed out "concentration" and "dilution" patterns based on nitrate-N concentration-discharge data, which is characterized by nitrate-N concentration peaks lagging the discharge peaks followed by rapid decline in the concentration as the streamflow hydrograph recedes. From Figures 5.12-5.13, it is noticed that the decline of nitrate-N concentrations during streamflow recession phases were not well represented, reflected by slower decrease comparing with the nitrate-N concentration observations. This could be attributed to improper calculation of flowpaths and their temporal variation (e.g. decrease of subsurface flow) because the nitrate-N is mainly driven by subsurface flow. The overestimation of IN concentrations during low flow conditions is probably caused by too simplified description of in-stream retention as discussed in Chapter 3 (Section 3.2).

For the catchment outlet (gauging station Hausneindorf) that is located in the lowland area, the observed IN concentrations show dampened dynamics compared with two upper-stream stations. During March-June of 2011, IN concentrations are underestimated due to under-prediction of discharge (Figures 5.12 and 5.13). In validation period (2012), IN concentrations are over-predicted under both BWSC and DAOC. The observed IN concentrations are lower for the same dates in validation period compared to calibration period with the mean observed nitrate-N concentration during calibration and validation period of 4.13 mg L^{-1} and 2.89 mg L^{-1}, respectively. The decrease of IN concentrations may be attributed to substantial decrease of discharge comparing 2012 with 2011 that leads to more N-uptake by plants and periphyton and slightly increase of denitrification. Boyacioglu et al. (2012) reported the large impact of future discharge changes attributed to climate change on regulating the magnitude, seasonal pattern and variability of nitrogen retention through scenario analysis. Model performances on IN concentration simulations derived from DAOC are found to be more robust between calibration and validation periods compared with those obtained from BWSC in terms of R^2 and MAE (Table 5.12).

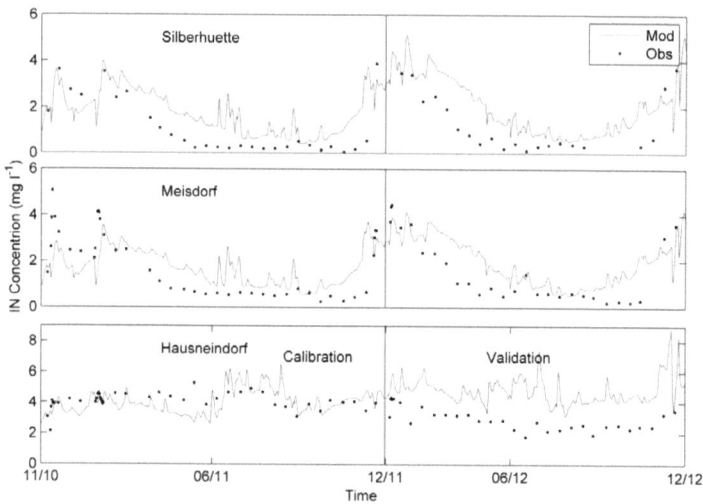

Figure 5.12 Observed and simulated IN concentrations in calibration (2010-2011) and validation (2012) periodsat three gauging stations using biweekly nitrate-N concentration measurements (BWSC).

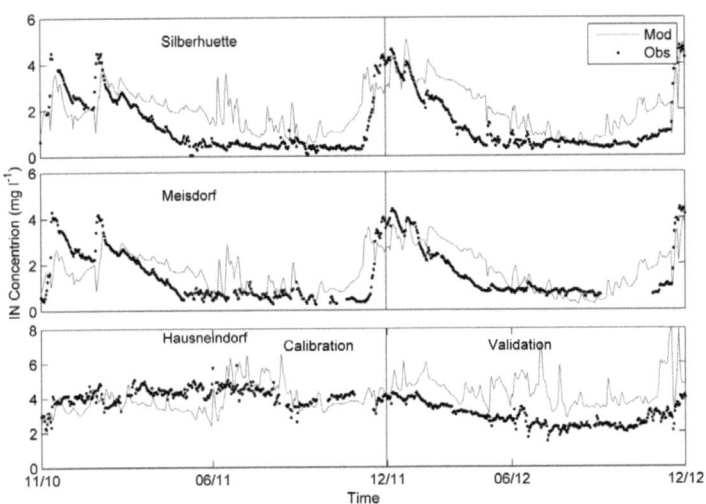

Figure 5.13 Observed and simulated IN concentrations in calibration (2010-2011) and validation (2012) periodsat three gauging stations using daily averages of nitrate-N concentration observations (DAOC).

Seasonal dynamics of daily IN loads are represented under calibration and validation againstBWSC and DAOC (Figures 5.14-5.15). Similar as IN concentrations, daily IN loads are characterized by high loads in peakflow events in winter and low loads during low flow conditions in summer. The under-prediction of daily IN loads in winter of 2010-2011 is mainly attributed to underestimations of discharges and IN concentrations. The daily IN loads are overestimated during validation period at the catchment outlet due to over-prediction of IN concentrations. The HYPE model have better and more robust performances between calibration and validation modes under DAOC compared with BWSC in terms of R^2, NSE, PBIAS and RSR at all three gauging stations (Table 5.12). At the catchment outlet (gauging station Hausneindorf), model performance during validation period under DAOC is not satisfactory attributed to overestimation of IN concentrations.

The measured annual IN loads were estimated to be 5.44 kg ha^{-1} yr^{-1} and 2.68 kg ha^{-1} yr^{-1} for 2011 and 2012, respectively using the approach explained in Chapter 2 (Section 2.5.2.1). Under BWSC, the predicted annual IN leaching loads for 2011 and 2012 are 5.81 kg ha^{-1} yr^{-1} and 4.66 kg ha^{-1} yr^{-1}, respectively. In comparison, the corresponding predicted annual IN leaching loads for 2011 and 2012 are 5.66 kg ha^{-1} yr^{-1} and 4.32 kg ha^{-1} yr^{-1}, respectively under DAOC. Therefore, the annual IN leaching loads are predicted with high accuracy for the calibration period (2011) under calibrations using both nitrate-N concentration sampling strategies (BWSC and DAOC). The annual IN leaching loads are over-predicted during validation period (2012) under both BWSC and DAOC due to overestimation of IN concentrations. However, annual IN loads predicted through calibration using daily averages of nitrate-N concentration observations are more accurate compared with that estimated under calibration against biweekly nitrate-N concentration measurements for both calibration (2011) and validation (2012) periods.

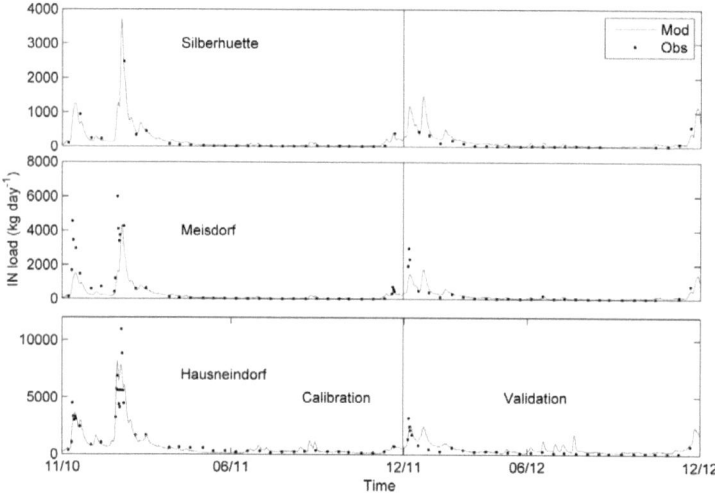

Figure 5.14 Observed and simulated daily IN loads in calibration (2010-2011) and validation (2012) periods at three gauging stations under biweekly nitrate-N concentration measurements (BWSC).

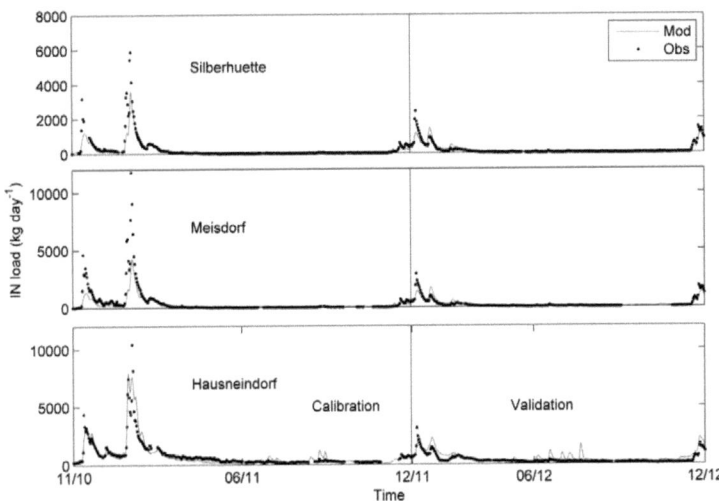

Figure 5.15 Observed and simulated daily IN loads in calibration (2010-2011) and validation (2012) periods at three gauging stations under daily averages of nitrate-N concentration observations (DAOC).

Table 5.12 Statistical model performances on simulations of IN concentration and daily IN loads in calibration (2011) and validation (2012) periods with parameter sets calibrated using PEST against BWSC (biweekly nitrate-N concentration measurements) and DAOC (daily averages of nitrate-N concentration observations).

Variable	Criterion	Calibration							Validation						
		Silberhuette		Meisdorf		Hausneindorf		Silberhuette		Meisdorf		Hausneindorf	
		BWSC	DAOC	BWSC	DAOC	BWSC	DAOC	BWSC	DAOC	BWSC	DAOC	BWSC	DAOC
IN concentration	R^2	0.58	0.36	0.18	0.30	0.06	0.05	0.61	0.67	0.65	0.60	0.15	0.17
	$MAE\ (mg\ L^{-1})$	0.845	0.973	0.988	0.722	0.580	0.743	1.02	0.891	0.842	0.685	1.90	1.51
Daily IN load	R^2	0.99	0.89	0.80	0.88	0.87	0.90	0.81	0.81	0.92	0.77	0.77	0.73
	NSE	0.96	0.78	0.47	0.61	0.85	0.87	0.70	0.79	0.72	0.76	0.70	0.33
	PBIAS	-2.57	-16.25	-51.14	-39.87	8.54	1.49	57.28	21.45	-20.38	8.99	36.23	61.61
	RSR	0.19	0.47	0.73	0.62	0.38	0.37	0.55	0.46	0.53	0.48	0.55	0.82

Predictive uncertainty

In implementing predictive analysis for IN concentration simulation using PEST, the key prediction was chosen as the high IN concentration observed on 10.05.2011 at catchment outlet (gauging station Hausneindorf) with measured value of 5.28 mg L^{-1}. Figure 16 presents the IN concentration prediction uncertainty estimated from calibration and predictive analysis using PEST under BWSC and DAOC. The predicted IN concentrations using the best optimized parameter set located inside the uncertainty ranges under both calibration and predictive analysis procedures (BWSC and DAOC). The uncertainty ranges obtained from calibration and predictive analysis using DAOC are much narrower than that derived from calibration and predictive analysis against BWSC. This indicates that calibration and predictive analysis using IN concentration observations of higher temporal resolution can improve prediction accuracy by decreasing predictive uncertainty range, which is attributed to better identified parameter set with increasing information content and constraints on nitrogen processes.

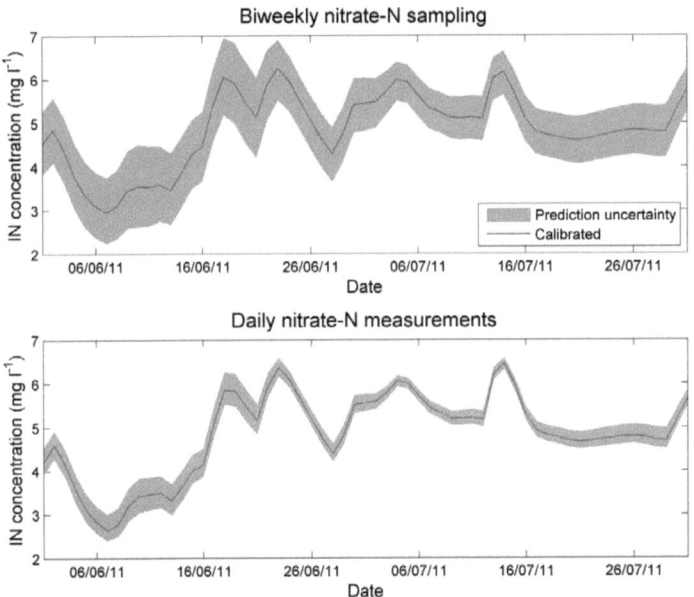

Figure 5.16 IN concentration prediction uncertainty at catchment outlet (gauging station Hausneindorf) for June-July of 2011estimated from calibration and predictive analysis using PEST against biweekly ni-

trate-N concentration measurements (BWSC) and daily averages of nitrate-N concentration observations (DAOC).

5.2.3 Calibration and predictive analysis using DREAM$_{(ZS)}$

Parameter identification

For calibrating nitrogen-process parameters and estimating IN concentration prediction uncertainty using DREAM$_{(ZS)}$, the same parameters and initial ranges as that selected in PEST calibration and predictive analysis run were defined (Table 4.3). The number of model runs was defined as 10000. Three Markov chains (N=3) were specified for the global search. Standard least squares (SLS) were chosen as likelihood function. The parameter posterior distributions derived from calibrations against BWSC and DAOC are shown in Figure 5.17. Model calibration results (including optimized values, mean values and standard deviation) derived from both calibration procedures (BWSC and DAOC) are given in Table 5.13. The large difference between parameter values calibrated based on BWSC and DAOC are mainly attributed to different number of nitrate-N observations (i.e. constraints) used for parameter identification. According to Figure 5.17, nitrogen-process parameters are better identified through calibration against DAOC compared to calibration against BWSC except for the parameter *wprod* due to its low sensitivity (Table 5.11). This is reflected by the narrower parameters' uncertainty ranges, sharp and peaked distributions estimated from calibration under DAOC (Blasone *et al.*, 2008). The smaller standard deviations of the calibrated parameters derived from calibration under DAOC compared to BWSC (except for parameter *wprod*) also indicates that the water quality parameters are better identified through calibration against IN concentration measurements of higher resolution. Sensitive parameters have similar optimized values estimated from calibrations under BWSC and DAOC, such as parameters *uptsoil102* and *denitw*. This indicates that the water quality parameters of the HYPE model are highly identifiable.

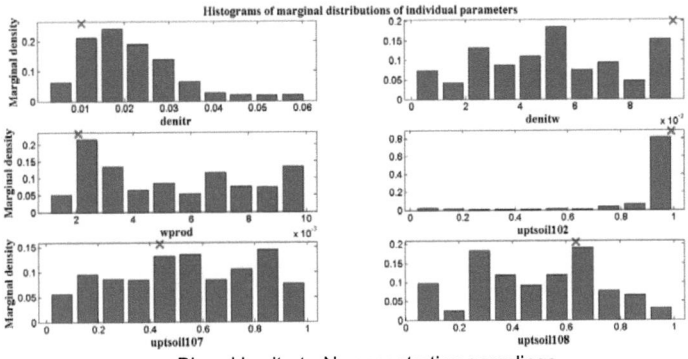

Biweekly nitrate-N concentration samplings

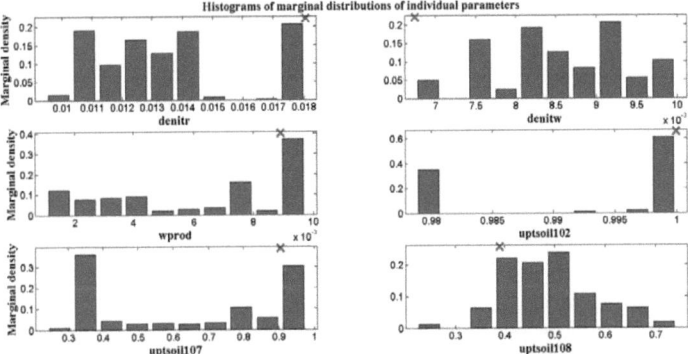

Daily nitrate-N concentration measurements

Figure 5.17 Histograms of marginal distributions of individual parameters derived from calibration against biweekly nitrate-N concentration measurements (BWSC) (top panel) and daily averages of nitrate-N concentration observations (DAOC) (bottom panel).

Table 5.13 Parameters calibration results from calibrations against biweekly nitrate-N concentration measurements (BWSC) and daily averages of nitrate-N concentration observations (DAOC).

Parameter	BWSC			DAOC		
	Optimized value	Mean value	Standard Dev	Optimized value	Mean value	Standard Dev
denitr	0.0102	0.0219	0.0114	0.0181	0.0136	0.0025
denitw	0.0096	0.0052	0.0027	0.0067	0.0085	0.0008
wprod	0.0021	0.0053	0.0027	0.0089	0.0062	0.0029
uptsoil102	0.99	0.90	0.20	1.00	0.99	0.009
uptsoil107	0.44	0.53	0.26	0.91	0.64	0.27
uptsoil108	0.64	0.49	0.24	0.39	0.49	0.09

Model performance

The statistical model performances on simulations of IN concentration and daily IN loads at three gauging stations (Silberhuette, Meisdorf and Hausneindorf) for both calibration (2010-2011) and validation (2012) modes with parameter sets calibrated using DREAM$_{(ZS)}$ against BWSC and DAOC are given in Table 5.14. Considering IN concentration simulations, model performances derived from calibration and validation under DAOC are better and more robust than those obtained from calibration and validation under BWSC in terms of MAE at all three gauging stations. Better and more robust model performances on daily IN loads simulations derived from DAOC at gauging stations Silberhuette and Meisdorf in terms of R^2, NSE and RSR compared to BWSC (Table 5.14). The HYPE model simulates daily IN loads at all three gauging stations satisfactorily using both calibrated parameter sets for calibration and validation modes.

Using the nitrogen-process parameters calibrated under BWSC, the estimated annual IN leaching loads were 5.96kg ha^{-1} yr^{-1} in calibration (2011) and 4.15 kg ha^{-1} yr^{-1} in validation (2012) periods. In contrary, the calculated annual IN leaching loads using the parameter set optimized againstDAOCwere 5.01 kg ha^{-1} yr^{-1} and 3.97 kg ha^{-1} yr^{-1} for calibration (2011) and validation (2012) periods, respectively. Referring to the measured annual IN leaching loads of 5.44 kg ha^{-1} yr^{-1} for 2011 and 2.68 kg ha^{-1} yr^{-1} for 2012, the annual IN leaching loads estimated through calibration against daily averages of nitrate-N concentration observations (DAOC) are more precise compared to that predicted through calibration against biweekly nitrate-N concentration measurements (BWSC).

Table 5.14 Statistical model performances on simulations of IN concentration and daily IN loads in calibration (2011) and validation (2012) periods with parameter sets calibrated using DREAM$_{(ZS)}$ against BWSC (biweekly nitrate-N concentration measurements) and DAOC (daily averages of nitrate-N concentration observations).

Variable	Criterion	Calibration						Validation					
		Silberhuette		Meisdorf		Hausneindorf		Silberhuette		Meisdorf		Hausneindorf	
		BWSC	DAOC	BWSC	DAOC	BWSC	DAOC	BWSC	DAOC	BWSC	DAOC	BWSC	DAOC
IN concentration	R^2	0.62	0.53	0.20	0.32	0.05	0.05	0.59	0.72	0.67	0.64	0.38	0.33
	$MAE\ (mg\ L^{-1})$	0.826	0.684	0.988	0.684	0.746	0.807	0.921	0.635	0.761	0.583	1.28	1.09
Daily IN load	R^2	0.98	0.86	0.75	0.85	0.78	0.76	0.76	0.80	0.92	0.76	0.76	0.76
	NSE	0.98	0.81	0.48	0.64	0.62	0.71	0.68	0.78	0.70	0.74	0.72	0.50
	PBIAS	5.63	-16.86	-50.49	-39.34	20.61	-5.06	46.68	8.22	-24.93	0.76	26.09	48.57
	RSR	0.14	0.44	0.72	0.60	0.62	0.54	0.56	0.47	0.55	0.51	0.53	0.71

Predictive uncertainty

The prediction uncertainty of IN concentration simulation at catchment outlet (gauging station Hausneindorf) for the calibration period (01.11.2010-31.12.2011) estimated from calibrations against BWSC and DAOC are shown in Figure 5.18. The grey ranges and black ranges stand for total uncertainty and parameter uncertainty, respectively. The measured IN concentrations are represented by red dots. As the amounts of IN concentration observations under these two different calibration procedures are different, the corresponding dynamics of estimated uncertainty ranges show different patterns. It is reflected by more detailed IN concentration dynamics estimated from DAOC compared to BWSC. Most importantly, the uncertainty ranges estimated from calibration against DAOC are much narrower than that estimated against BWSC in terms of both parameter uncertainty and total uncertainty. This can be explained by the fact that daily averages of nitrate-N concentration measurements (DAOC) increase constraints on nitrogen transport and transformation processes because various weather, hydrological and nitrogen conditions are included. As a result, the nitrogen-process parameters were identified with higher certainty and the IN concentration prediction uncertainty were lower compared to calibration against biweekly nitrate-N concentration measurements (BWSC).

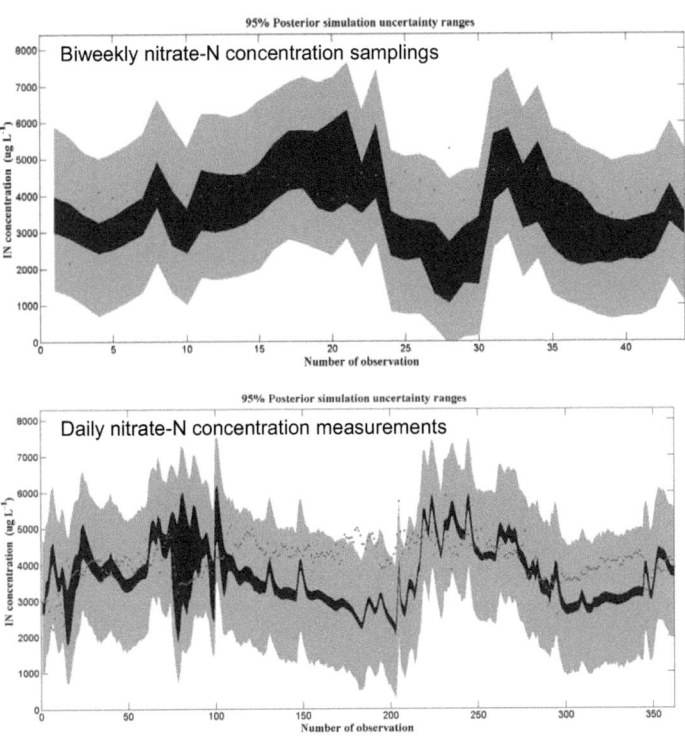

Figure 5.18 IN concentration prediction uncertainty (parameter uncertainty (black area) and total uncertainty (grey area)) together with observed IN concentrations (red dots) estimated from calibration against biweekly nitrate-N concentration measurements (BWSC) (top panel) and daily averages of nitrate-N concentration observations (DAOC) (bottom panel).

5.3 Comparison of PEST and DREAM$_{(ZS)}$

In this study both the local search approach PEST and global search approach DREAM$_{(ZS)}$ were applied for hydrological water quality parameter identification and predictive analysis on the HYPE model. The optimized parameter values estimated from PEST and DREAM$_{(ZS)}$ are similar when the same initial parameter settings are defined (See Tables 5.1 and 5.4 for hydrological parameter identification; Tables 5.6 and 5.8 for nitrogen-process parameter identification). The model performances with best parameter sets calibrated using both approaches (PEST and DREAM$_{(ZS)}$) are also comparable (See Tables 5.2 and 5.5 for discharge simulations; Tables 5.7 and 5.9 for IN parameter identification). This indicates the hydrological water quality parameters of the HYPE model are highly identifiable and both approaches are efficient in parameter calibration. Considering the capability of estimating predictive uncertainty, both approaches are capable to reveal that increasing spatial and temporal resolution of discharge and IN concentration measurements decrease prediction uncertainty.

Using $DREAM_{(ZS)}$ it is capable to evolve the parameter density functions (i.e. parameter posterior distributions) following Bayesian inference from user-defined initial uniform-distributed parameter space, which cannot be done with PEST. Based on the model simulation results using all posterior parameter combinations derived from $DREAM_{(ZS)}$, prediction uncertainty ranges (e.g. 95% ranges) are projected. While using PEST for predictive analysis, the model is run in "dual calibration" mode; the selection of key prediction and definition of calibration threshold are subjective because they are dependent on the goodness of model calibration and obtained minimum objective function (See Section 2.4.1.3 and Appendix 2C). Moreover, the parameter calibration and predictive uncertainty results using PEST are highly dependent on the choice of parameters to be calibrated and definition of initial values and ranges attributed to its local search nature (Bahremand and De Smedt, 2010). The sole limitation of $DREAM_{(ZS)}$ in parameter calibration and uncertainty analysis is related to its higher computational cost compared with PEST. However, with the development of computing power and introduction of parallel computing, model calibration and uncertainty analysis can be realized using $DREAM_{(ZS)}$ within acceptable time (Laloy and Vrugt, 2012). From this study, it is concluded that PEST is more efficient for hydrological water quality calibration than $DREAM_{(ZS)}$ related to its lower computational cost; while $DREAM_{(ZS)}$ is more appropriate for model uncertainty analysis due to its global search nature and Bayesian inference.

Chapter 6: Conclusions

This dissertation has five goals: (i) to evaluate the capability of the HYPE model to simulate streamflow and streamwater IN concentration and IN loads at catchments with different climatic, physiographic, hydrological and water quality conditions; (ii) to test different calibration strategies (step-wise calibration and multi-objective calibration) on optimization of integrated hydrological water quality models; (iii) to investigate effects of spatial resolution of calibration data (discharge and IN concentration measurements) on model identification and prediction accuracy; (iv)to investigate effects of temporal resolution of calibration data (IN concentration measurements) on model identification and prediction accuracy; (v) to evaluate the efficiency of different calibration and uncertainty analysis approaches (PEST and DREAM$_{(ZS)}$) on model identification.

For the first goal, the HYPE model was set up at two mesoscale catchments (Selke and Weida) to simulate streamflow and streamwater IN concentration and IN loads. These two catchments show different climate (precipitation) patterns, land use characteristics, hydrological andwater qualityconditions. The multi-site and multi-objective calibration method was employed using PEST for parameter identification. Through sensitivity analysis using PEST, the most sensitive hydrological and nitrogen processes are evapotranspiration, plant uptake and denitrification, respectively. The HYPE model is found to be able to represent dynamics and balances of streamflow and IN loads at both catchments for calibration and validation modes. IN concentration and daily IN loads show a proportional relationship with discharge, indicating that IN leaching is mainly controlled by discharge in these managed catchments. This is consistent with former findings. The change of model performance across different parts of the Selke catchment in terms of water balance implies that the simplification and assumption of the HYPE model for the evapotranspiration process can not represent the effects of spatial variability of elevation and temperature. Overestimation of IN concentration during low flow conditions indicates that IN retention (e.g. denitrification in the soil and in the river system) processes are not properly described in the model. The effects of fine sediments on nitrogen retention in the river (e.g. adsorbtion/desorption) need to be considered; the influences of soil moisture on denitrification in the soil and the effects of river morphology on in-stream retention need to be further considered. The HYPE model is able to correctly predict IN losses from agricultural land with strongly variable runoff and share of arable land. This was also true for extreme weather conditions like the extreme period of 2002-2004. Although model improvements seem to be necessary for instream IN retention, it can be concluded that the process-based model approach employed in the HYPE model provides realistic estimates of IN yields. Therefore, the HYPE model can be used as a reliable decision-making tool for the environmental protection agencies.

For the second goal, PEST was combined with the HYPE model to implement and compare the stepwise calibration (i.e. calibrating hydrological parameters first and then calibrating water quality parameters) in combination with the multi-objective calibration (i.e. calibrating hydrological and water quality parameters simultaneously).The multi-objective calibration was found to be more appropriate for parameter identification in integrated hydrological water quality modeling because of more robust model performances on streamflow and IN simulations between calibration and validation periods. This is explained by the fact that hydrological water quality processes are correlated and multi-objective calibration enables to increase constraints and make most of the different types of measurements. This verifies that multi-objective calibration is more reasonable and appropriate for parameter calibration in integrated hydrological water quality modeling.

For the third goal, both PEST and DREAM$_{(ZS)}$ were combined with the HYPE model to implement hydrological water quality parameter calibration and predictive analysis at Selke catchment. Through comparing posterior parameter uncertainty, model performance and prediction uncertainty derived from single-site and multi-site calibrations, it is found that multi-site calibration improves model performance on IN simulation at internal sites and reduces hydrological water quality parameters' uncertainty and prediction uncertainty compared to single-site calibration. This indicates the importance of observations from internal sites for spatially distributed predictions.

For the fourth goal, Model performance, parameter and prediction uncertainty obtained from calibration and validation against biweekly nitrate-N concentration measurements and daily averages of nitrate-N concentration observations were compared. Calibration using daily averages of nitrate-N concentration observations are found to improve model identification by decreasing parameters' posterior uncertainty ranges and prediction uncertainty, as well as improving model performance on IN concentration and IN loads simulations compared to calibration using biweekly nitrate-N concentration measurements. This verifies the importance of high resolution water quality measurements for model identification and nutrient load calculation.

For the fifth goal, both PEST and DREAM$_{(ZS)}$ are found to be efficient for hydrological water quality parameter calibration. However, DREAM$_{(ZS)}$ is more reasonable and appropriate for parameter identification and uncertainty analysis due to its Bayesian inference and global search nature.

References

Abbaspour KC, Yang J, Maximov I, Siber R, Bogner K, Mieleitner J, Zobrist J, Srinivasan, R. 2007. Modelling hydrology and water quality in the pre-alpine/alpine Thur watershed using SWAT. *Journal of Hydrology* 333(2-4): 413-430.

Andersen HE, Kronvang B, Larsen SE, Hoffmann CC, Jensen TS, Rasmussen EK. 2006. Climate-change impacts on hydrology and nutrients in a Danish lowland river basin. *Science of the Total Environment* 365: 223-237. Doi: 10.1016/j.scitotenv.2006.02.036.

Andersson L, Rosberg J, Pers BC, Olsson J, Arheimer B. 2005. Estimating catchment nutrient flow with the HBV-NP model: sensitivity to the input data. *Ambio* 34: 439-447.

Andréassian V, Perrin C, Michel C, Usart-Sanchez I, Lavabre J. 2001.Impact of imperfect rainfall knowledge on the efficiency and the parameters of watershed models. *Journal of hydrology* 250: 206-223.

Arheimer B, Brandt M. 1998.Modelling nitrogen transport and retention in the catchments of southern Sweden. *Ambio* 27: 471-480.

Arheimer B, Löwgren M, Pers BC, Rosberg J. 2005. Integrated catchment modeling for nutrient reduction: scenarios showing impacts, potential, and cost of measures. *Ambio* 34: 513-520.

Arnell NW. 1999. The effect of climate change on hydrological regimes in Europe: a continental perspective. *Global Environmental Change* 9: 5-23.

Arnold JG, Srinivasan R, Muttiah RS, Williams JR. 1998. Large area hydrological modeling and assessment part I: model development. *Journal of the American Water Resources Association* 34: 73-89. Doi: 10.1111/j.1752-1688.1998.tb05961.x.

Askegaard M, Olesen JE, Kristensen K. 2005. Nitrate leaching from organic agriculture crop rotations: Effects of location, manure and catch crop. *Soil Use and Management* 21: 181-188. Doi: 10.1111/j.1475-2743.2005.tb00123.x.

Bahremand A, De Smedt F. 2010. Predictive Analysis and Simulation Uncertainty of a Distributed Hydrological Model.*Water Resources Management* 24: 2869-2880. Doi: 10.1007/s11269-010-9584-1.

Balin D, Lee H, Rode M. 2010. Is point uncertain rainfall likely to have a great impact on distributed complex hydrological modeling? *Water Resources Research* W11520, Doi:10.1029/2009WR007848.

Basu NB, Destouni G, Jawitz JW, Thompson SE, Loukinova NV, Darracq A, Zanardo S, Yaeger M, Sivapalan M, Rinaldo A, Rao PSC. 2010. Nutrient loads exported from managed catchments reveal emergent biogeochemical stationarity. *Geophysical Research Letters* 37: L23404. Doi: 10.1029/2010GL045168.

Basu NB, Thompson SE, Rao PSC. 2011. Hydrologic and biogeochemical functioning of intensively managed catchments: A synthesis of top-down analyses. *Water Resources Research* 47: W00J15. Doi: 10.1029/2011WR010800.

Becker A, Braun P. 1999. Disaggregation, aggregation and spatial scaling in hydrological modelling. *Journal of hydrology* 217: 239-252.

Becker A, Mcdonnell JJ. 1998. Topographical and ecological controls of runoff generation and lateral flows in mountain catchments. Hydrology, Water Resources and Ecology in Headwaters (*Proceedings of the Head-Water '98 Conference* held at Meran/Merano, Italy, April 1998). IAHS Publ. no. 248.

Bergström S. 1976. Development and application of a conceptual runoff model for Scandinavian catchments. SMHI Reports RHO, No. 7, Norrköping.

Benke KK, Lowell KE, Hamilton AJ. 2008. Parameter uncertainty, sensitivity analysis and prediction error in a water-balance hydrological model. *Mathematical and Computer Modelling* 47: 1134-1149.

Beven KJ. 2007. Towards integrated environmental models of everywhere: uncertainty, data and modelling as a learning process. *Hydrology and Earth System Sciences* 11: 460-467.

Beven KJ. 1993. Prophecy, reality and uncertainty in distributed hydrological modelling. *Advances in Water Resources* 16: 41-51.

Beven KJ, Freer J. 2001. Equifinality, data assimilation, and uncertainty estimation in mechanistic modelling of complex environmental systems using the GLUE methodology. *Journal of hydrology* 249: 11-29. Doi: 10.1016/S0022-1694(01)00421-8.

Beven KJ, Binley AM. 1992. The future of distributed models: model calibration and predictive uncertainty. *Hydrological Processes* 6: 279-298.

Beven KJ, Smith PJ, Freer JE. 2008. So just why would a modeller choose to be incoherent? *Journal of Hydrology* 354(1-4): 15-32.

Bicknell BR, Imhoff JC, Kittle JLJr, Jobes TH, Donigian ASJr. 2001. *HSPF, Version 12, User's Manual*. U.S. Environmental Protection Agency, Athens, GA.

Birgand F, Skaggs RW, Chescheir GM, Gilliam JW. 2007. Nitrogen Removal in Streams of Agricultural Catchments—A Literature Review. *Critical Reviews in Environmental Science and Technology* 37: 381-487.

Blasone RS, Madsen H, Rosbjerg D. 2008. Uncertainty assessment of integrated distributed hydrological models using GLUE with Markov chain Monte Carlo sampling. *Journal of Hydrology* 353(1-2): 18-32.

Borah DK, Bera M. 2004. Watershed-scale hydrologic and nonpoint-source pollution models: review of applications. *Transactions of the ASAE* 47: 789–803.

Bosch NS. 2008. The influence of impoundments on riverine nutrient transport: An evaluation using the Soil and Water Assessment Tool. *Journal of Hydrology* 355(1-4): 131-147.

Bongartz K. 2004. Abschlußbericht der FSU Jena zum Forschungsprojekt "Integriertes Flussgebietsmanagement am Beispiel der Saale" (in German). Institut für Geographie, Friedrich-Schiller-Universität Jena, Germany.

Bouraoui F, Grizzetti B, Granlund K, Rekolainen S, Bidoglio G. 2004. Impact of climate change on the water cycle and nutrient losses in a Finnish catchment. Climatic Change 66: 109-126. Doi: 10.1023/B:CLIM.0000043147.09365.e3.

Boyacioglu H, Vetter T, Krysanova V, Rode M. 2011. Modeling the impacts of climate change on nitrogen retention in a 4th order stream. *Climatic Change* 113(3-4): 981-999.

Boyle DP, Gupta HV, Sorooshian S. 2000. Toward improved calibration of hydrologic models: Combining the strengths of manual and automatic methods. *Water Resources Research* 36: 3663-3674.

Böhme F, Merbach I, Weigel A, Russow R. 2003. Effect of crop type and crop growth on atmospheric nitrogen deposition. *Journal of Plant Nutrition and Soil Science* 166: 601-605. DOI: 10.1002/jpln.200321112.

Butts MB, Payne JT, Kristensen M, Madsen H. 2004. An evaluation of the impact of model structure on hydrological modelling uncertainty for streamflow simulation. *Journal of Hydrology* 298(1-4): 242-266.

BMU-Federal Ministry for the Environmen, Nature Conservation and Nuclear Safty 2005: Water Framework Directive - Summary to River Basin District Analysis 2004 in Germany, 68 p.

Cao W, Bowden WB, Davie T, Fenemor A. 2006. Multi-variable and multi-site calibration and validation of SWAT in a large mountainous catchment with high spatial variability. *Hydrological Processes* 20(5): 1057-1073.

Chen X, Cheng Q, Chen XD, Smettem K, Xu CX. 2010. Simulating the integrated effects of topography and soil properties on runoff generation in hilly forested catchments, South China. *Hydrological Processes* 24: 714-725. Doi: 10.1002/hyp.7509.

Cherry KA, Shepherd M, Withers PJ, Mooney SJ. 2008. Assessing the effectiveness of actions to mitigate nutrient loss from agriculture: a review of methods. *Science of the Total Environment* 406(1-2): 1-23.

Chu TW, Shirmohammadi A, Montas H, Sadeghi A. 2004.Evaluation of the SWAT model's sediment and nutrient components in the piedmont physiographic region of Maryland.*Transactions of the ASAE* 47: 1523–1538.

Creed IF, Band LE, Foster NW, Morrison IK, Nicolson JA, Semkin RS, Jeffries DS. 1996. Regulation of Nitrate-N Release from Temperate Forests: A Test of the N Flushing Hypothesis. *Water Resources Research* 32: 3337-3354. Doi: 10.1029/96WR02399.

Cuo L, Giambelluca TW, Ziegler AD. 2011. Lumped parameter sensitivity analysis of a distributed hydrological model within tropical and temperate catchments. *Hydrological Processes* 25: 2405-2421. Doi: 10.1002/hyp.8017.

Chahinian N, Tournoud, MG, Perrin JL, Picot B. 2011. Flow and nutrient transport in intermittent rivers: a modelling case-study on the Vène River using SWAT 2005. *Hydrological Sciences Journal* 56: 268-287. Doi: 10.1080/02626667.2011.559328.

De Klein JJM, Koelmans AA. 2011. Quantifying seasonal export and retention of nutrients in West European lowland rivers at catchment scale. *Hydrological Processes* 25: 2102-2111. Doi: 10.1002/hyp.7964.

De Wit MJM, Pebesma EJ. 2001. Nutrient fluxes at the river basin scale. II: the balance between data availability and model complexity. *Hydrological Processes* 15(5): 761-775.

Dekker SC, Vrugt JA, Elkington RJ. 2012. Significant variation in vegetation characteristics and dynamics from ecohydrological optimality of net carbon profit. *Ecohydrology* 5: 1-18.

Dodds WK. 2006.Eutrophication and trophic state in rivers and streams.*Limnol.Oceanogr.* 51(1,par2) 671-680.

Doherty J. 2005. PEST: model independent parameter estimation, user manual, 5th edn. Watermark Numerical Computing, Brisbane.

Doherty J, Johnston JM. 2003. Methodology for calibration and predictive analysis of a watershed model. *Journal of the American Water Resources Association (JAWRA)* 39(2): 251-265.

Drewry JJ, Newham LTH, Greene RSB, Jakeman AJ, Croke BFW. 2006. A review of nitrogen and phosphorus export to waterways: context for catchment modelling. *Marine and Freshwater Research* 57: 757-774.

Duan Q, Gupta VK, Sorooshian S. 1992. Effective and efficient global optimization for conceptual rainfall-runoff models. *Water Resources Research* 19: 260-268.

EC, 2000.Directive 2000/60/EC of the European Parliament and of the Council of 23October 2000 establishing a framework for Community action in the field of water policy. *Official Journalof the European Communities* L327.

European commission report.Implementation of the nitrates directive, COM (2002) 407.

Faurès JM, Goodrich DC, Woolhiser DA, Sorooshian S. 1995. Impact of small-scale spatial rainfall variability on runoff. *Journal of hydrology* 173: 309-326.

Feyen L, Vázquez R, Christiaens K, Sels O, Feyen J. 2000. Application of a distributed physically-based hydrological model to a medium size catchment. *Hydrology and Earth System Sciences* 4(1): 47-63.

Fink M. 2004. Regionale Modellierung der Wasser- und Stickstoff- dynamik als Entscheidungsunterstützung für die Reduktion des N-Eintrags am Bespiel des Trinkwassertalsperrensystems Weida-Zeulenroda, Thüringen, Dissertation Jena Friedrich- Schiller-Universität, Chemisch- Geowissenschaftliche Fakultät, Jena, Germany (in German).

FitzHugh TW, Mackay DS. 2000. Impacts of input parameter spatial aggregation on an agricultural nonpoint source pollution model. *Journal of hydrology* 236: 35-53.

Flügel WA. 1995. Delineating hydrological response units by geographical information system analyses for regional hydrological modelling using PRMS/MMS in the drainage basin of the River Bröl, Germany. *Hydrological Processes* 9: 423-436.

Gan TY, Dlamini EM, Biftu GF. 1997. Effects of model complexity and structure, data quality, and objective functions on hydrologic modeling. *Journal of Hydrology* 192: 81-103.

Gan TY, Biftu GF. 1996. Automatic Calibration of Conceptual Rainfall-Runoff Models: Optimization Algorithms, Catchment Conditions, and Model Structure. *Water Resources Research* 32: 3513-3524.

Gardner KK, McGlynn BL, Marshall LA. 2011. Quantifying watershed sensitivity to spatially variable N loading and the relative importance of watershed N retention mechanisms. *Water Resources Research* 47.

Gelman A, Rubin DB. 1992. Inference from iterative simulation using multiple sequence. *Statistical Science*, 7: 457-511.

Glavan M, White S, Holman IP. 2011. Evaluation of river water quality simulations at a daily time step - experience with SWAT in the Axe catchment, UK. *CLEAN - Soil Air Water* 39: 43-54. Doi: 10.1002/clen.200900298.

Grizzetti B, Bouraoui F, Granlund K, Rekolainen S, Bidoglio G. 2003. Modelling diffuse emission and retention of nutrients in the Vantaanjoki watershed (Finland) using the SWAT model. *Ecological Modelling* 169: 25-38. Doi: 10.1016/S0304-3800(03)00198-4.

Groenendijk P, Kroes JG. 1999. Modelling the nitrogen and phosphorus leaching to groundwater and surface water with ANIMO 3.5. Winand Starting Centre.Wageningen (The Netherlands), Report 144.

Gupta HV, Sorooshian S, Yapo PO. 1999. Status of automatic calibration for hydrologic models: comparison with multilevel expert calibration. *Journal of Hydrologic Engineering* 4: 135-143. Doi: 10.1061/(ASCE)1084-0699.

Haario H, Laine M, Mira A, Saksman E. 2006. DRAM: Efficient adaptive MCMC. *Statistics and Computing* 16: 339-354.

Haario H, Saksman E, Tamminen J. 2001. An adaptive Metropolis algorithm. Bernoulli, 7: 223-242.

Haberlandt U, Ebner von Eschenbach AD, Buchwald I. 2008. A space-time hybrid hourly rainfall model for derived flood frequency analysis. *Hydrology and Earth System Sciences* 12: 1353-1367. Doi: 10.5194/hess-12-1353-2008.

Hansen B, Alrøe HF, Kristensen ES. 2001. Approaches to assess the environmental impact of organic farming with particular regard to Denmark. *Agriculture Ecosystems and Environment* 83: 11-25.

Heathwaite L. 1995. Sources of eutrophication: hydrological pathways of catchment nutrient export. Man's Influence on Freshwater Ecosystems and Water Use (*Proceedings of a Boulder Symposium*, July 1995). IAHS Publ. no. 230.

Helsel DR, Hirsch RM. 1992. Statistical methods in water resources. Elsevier, New York, P 126.

Hesser FB, Franko U, Rode M. 2010. Spatial distributed lateral nitrate transport at the catchment scale. *Journal of Environmental Quality* 39: 193-203. Doi: 10.2134/jeq2009.0031.

Hunukumbura PB, Tachikawa Y, Shiiba M. 2012. Distributed hydrological model transferability across basins with different hydro-climatic characteristics. *Hydrological Processes* 26: 793-808. Doi: 10.1002/hyp.8294.

Jain S, Sudheer K. 2008. Fitting of Hydrologic Models: A Close Look at the Nash–Sutcliffe Index. *Journal of Hydrologic Engineering* 13: 981-986. Doi: 10.1061/(ASCE)1084-0699(2008)13:10(981).

Jiang S, Rode M. 2012. Modeling water flow and nutrient losses (Nitrogen, Phosphorus) at a nested mesoscale catchment, Germany. International Congress on Environmental Modelling and Software(iEMSs), Managing Resources of a Limited Planet, Sixth Biennial Meeting, Leipzig, Germany.

Jiang S, Jomaa S, Rode M. 2014. Modelling inorganic nitrogen emissions at a nested mesoscale catchment in central Germany. *Ecohydrology*. Doi: 10.1002/eco.1462.

Johnsson H, Bergström L, Jansson PE. 1987. Simulated nitrogen dynamics and losses in a layered agricultural soil. *Agriculture, Ecosystems and Environment* 18: 333-356.

Lagzdins A, Jansons V, Sudars R, Abramenko K. 2012. Scale issues for assessment of nutrient leaching from agricultural land in Latvia. *Hydrology Research* 43: 383-399. Doi:10.2166/nh.2012.122.

Laloy E, Vrugt JA. 2012. High-dimensional posterior exploration of hydrologic models using multiple-try DREAM$_{(ZS)}$ and high-performance computing. *Water Resources Research* 48, W01526, Doi: 10.1029/2011WR010608.

Lam QD, Schmalz B, Fohrer N. 2012. Assessing the spatial and temporal variations of water quality in lowland areas, Northern Germany. *Journal of Hydrology* 438-439: 137-147.

Landesbericht, 2005. Landesbericht über die Bestandsaufnahme der Gewässer nach Artikel 5. EG-WRRL, LandSachsen-Anhalt (C-Bericht).

Li H, Sivapalan M, Tian F, Liu D. 2010. Water and nutrient balances in a large tile-drained agricultural catchment: a distributed modeling study. *Hydrology and Earth System Sciences* 14: 2259-2275. Doi: 10.5194/hessd-7-3931-2010.

Lin ZL, Radcliffe DE. 2006. Automatic Calibration and Predictive Uncertainty Analysis of a Semidistributed Watershed Model. *Vadose Zone Journal* 5: 248-260.

Lord EI, Anthony SG, Goodlass G. 2002. Agricultural nitrogen balance and water quality in the UK. *Soil Use and Management* 18: 363-369.

Lu S, Kayastha N, Thodsen H, van Griensven A, Andersen HE. 2011. Multiobjective calibration for comparing channel sediment routing models in the Soil and Water Assessment Tool. *Journal of Environmental Quality (special section)*. Doi: 10.2134/jeq2011.0364.

Littlewood IG. 1995. Hydrological regimes, asmpling strategies, and assessment of errors in mass load estimates for united kindom rivers. *Environment International* 21: 211-220.

Lindström G, Bishop K, Ottosson-Löfvenius M. 2002.Soil frost and runoff at Svartberget, northern Sweden-measurements and model analysis. *Hydrological Processes* 16: 3379-3392.

Lindström G, Johansson B, Persson M, Gardelin M, Bergström S. 1997. Development and test of the distributed HBV-96 hydrological model. *Journal of Hydrology* 201: 272-288. Doi: 10.1016/S0022-1694(97)00041-3.

Lindström G, Pers C, Rosberg J, Strömqvist J, Arheimer B. 2010. Development and testing of the HYPE (Hydrological Predictions for the Environment) water quality model for different spatial scales. *Hydrology Research* 41: 295-319. Doi: 10.2166/nh.2010.007.

Lindström G, Rosberg J, Arheimer B. 2005. Parameter precision in the HBV-NP model and impacts on nitrogen scenario simulations in the Rönneå River, Southern Sweden. *Ambio* 34: 533-537.

Liu F, Parmenter R, Brooks PD, Conklin MH, Bales RC. 2008. Seasonal and interannual variation of streamflow pathways and biogeochemical implications in semi-arid, forested catchments in Valles Caldera, New Mexico. *Ecohydrology* 1: 239-252. Doi: 10.1002/eco.22.

Kavetski D, Kuczera G, Franks SW. 2006. Bayesian analysis of input uncertainty in hydrological modeling: 2. Application. *Water Resources Research* 42. Doi: 10.1029/2005wr004376.

Kistner I. 2007. Anwendung des Modells ANIMO zur Simulation des gelösten Phosphors im Oberflächenabfluss auf der Feldskala und der Phosphorverfügbarkeit im Oberboden auf der Einzugsgebietsskala. Ph.D. Thesis: 17-20.

Krysanova V, Hattermann F, Wechsung F. 2005. Development of the ecohydrological model SWIM for regional impact studies and vulnerability assessment. *Hydrological Processes* 19: 763-783. Doi: 10.1002/hyp.5619.

Kyllmar K, Carlsson C, Gustafson A, Ulén B, Johnsson H. 2006. Nutrient discharge from small agricultural catchments in Sweden: Characterisation and trends. *Agriculture Ecosystems and Environment* 115: 15-26. Doi: 10.1016/j.agee.2005.12.004.

Kistner I, Ollesch G, Meissner R., Rode M. 2013. Spatial-temporal dynamics of water soluble phosphorus in the topsoil of a low mountain range catchment. *Agriculture, Ecosystems & Environment* 176: 24-38.

Madsen H. 2003. Parameter estimation in distributed hydrological catchment modelling using automatic calibration with multiple objectives. *Advances in Water Resources* 26: 205-216.

Madsen H. 2000. Automatic calibration of a conceptual rainfall–runoff model using multiple objectives. *Journal of hydrology* 235: 276-288.

McCuen R, Knight Z, Cutter A. 2006. Evaluation of the Nash–Sutcliffe Efficiency Index. *Journal of Hydrologic Engineering* 11: 597–602. Doi: 10.1061/(ASCE)1084-0699(2006)11:6(597).

Miltner RJ, Rankin ET. 1998. Primary nutrients and the biotic integrity of rivers and streams. *Freshwater Biology* 40: 145-158.

Moriasi DN, Arnold JG, Liew MWV, Bingner RL, Harmel RD, Veith TL. 2007. Model evaluation guidelines for systematic quantification of accuracy in watershed simulations. *Transactions of the ASABE* 50: 885-900.

Moussa R, Chahinian N, Bocquillon C. 2007. Distributed hydrological modelling of a Mediterranean mountainous catchment - Model construction and multi-site validation. *Journal of Hydrology* 337: 35-51.

Mousavi SJ, Abbaspour KC, Kamali B, Amini M, Yang H. 2012. Uncertainty-based automatic calibration of HEC-HMS model using sequential uncertainty fitting approach. *Journal of Hydroinformatics* 14: 286.

Nafees Ahmad HM, Sinclair A, Jamieson R, Madani A, Hebb D, Havard P, Yiridoe EK. 2011. Modeling sediment and nitrogen export from a rural watershed in eastern Canada using the soil and water assessment tool. *Journal of Environtal Quality* 40(4): 1182-94.

Nash JE, Suttcliffe JV. 1970. River flow forecasting through conceptual models part I- A discussion of principles. *Journal of Hydrology* 10: 282-290. Doi: 10.1016/0022-1694(70)90255-6.

Nelder JA, Mead R. 1965. A simplex method for function minimization.*The Computer Journal* 7: 308-313.

Onderka M, Wrede S, Rodný M, Pfister L, Hoffmann L, Krein A. 2012. Hydrogeologic and landscape controls of dissolved inorganic nitrogen (DIN) and dissolved silica (DSi) fluxes in heterogeneous catchments. *Journal of Hydrology* 450-451: 36-47. Doi: 10.1016/j.jhydrol.2012.05.035.

Perrin C, Michel C, Andréassian V. 2001. Does a large number of parameters enhance model performance? Comparative assessment of common catchment model structures on 429 catchments. *Journal of Hydrology* 242: 275-301.

Refsgaard JC. 1997. Parameterisation, calibration and validation of distributed hydrological models. *Journal of hydrology* 198: 69-97.

Refsgaard JC, Henriksen HJ. 2004. Modelling guidelines—terminology and guiding principles. *Advances in Water Resources* 27(1): 71-82.

Refsgaard JC, Knudsen J. 1996. Operational Validation and Intercomparison of Different Types of Hydrological Models. *Water Resources Research* 32(7): 2189-2202.

Refsgaard JC, van der Sluijs JP, Brown J, van der Keur P. 2006. A framework for dealing with uncertainty due to model structure error. *Advances in Water Resources* 29(11): 1586-1597.

Rode M, Arhonditsis G, Balin D, Kebede T, Krysanova V, van Griensven A, van der Zee SEATM. 2010. New challenges in integrated water quality modelling. *Hydrological Processes* 24: 3447-3461. Doi: 10.1002/hyp.7766.

Rode M, Klauer B, Krause P, Lindenschmidt KE. 2002. Integrated river basin management: a new ecologically-based modelling approach. *Ecohydrology & Hydrobiology* 2: 171-179.

Rode M, Klauer B, Petry D, Volk M, Wenk G, Wagenschein D. 2008. Integrated nutrient transport modelling with respect to the implementation of the European WFD: The Weiße Elster Case Study, Germany. *Water SA* 34: 490-496.

Rode M, Lindenschmidt KE.2001.Distributed sediment and phosphorus transport modeling on a medium sized catchment in central Germany. *Physics and Chemistry of the Earth (B)* 26: 635-640.

Rode M, Suhr U. 2007 a. Uncertainties in selected river water quality data. *Hydrol. Earth Syst. Sci.* 11: 863–874.

Rode M, Suhr U, Wriedt G. 2007 b. Multi-objective calibration of a river water quality model–Information content of calibration data. *Ecological Modelling* 204: 129-142. Doi: 10.1016/j.ecolmodel.2006.12.037.

Rode M, Thiel E, Franko U, Wenk G, Hesser F. 2009. Impact of selected agricultural management options on the reduction of nitrogen loads in three representative meso scale catchments in Central Germany. *Science of the Total Environment* 407: 3459-3472. Doi: 10.1016/j.scitotenv.2009.01.053.

Scharnagl B, Vrugt JA, Vereecken H, Herbst M. 2010. Information content of incubation experiments for inverse estimation of pools in the Rothamsted carbon model: a Bayesian perspective. *Biogeosciences* 7: 763-776.

Schoups G, Vrugt JA. 2010. A formal likelihood function for parameter and predictive inference of hydrologic models with correlated, heteroscedastic, and non-Gaussian errors. *Water Resources Research* 46(10).

Seneviratne SI, Lehner I, Gurtz J, Teuling AJ, Lang H, Moser U, Grebner D, Lucas M, Schroff K Vitvar T, Zappa M. 2012. Swiss prealpine Rietholzbach research catchment and lysimeter: 32 year time series and 2003 drought event. *Water Resources Research* 48: W06526. Doi: 10.1029/2011WR011749.

Singh J, Knapp HV, Arnold JG, Demissie M. 2005. Hydrological modelling of the iroquois river watershed using HSPF and SWAT. *Journal of the American Water Resources Association (JAWRA)* 41: 343-360.

Shrestha RR, Bárdossy A, Rode M. 2007. A hybrid deterministic–fuzzy rule based model for catchment scale nitrate dynamics. *Journal of Hydrology* 342: 143-156. Doi: 10.1016/j.jhydrol.2007.05.020.

Shrestha RR, Dibike YB, Prowse TD. 2012. Modeling Climate Change impacts on hydrology and nutrient loading in the upper Assiniboine catchment. *Journal of the American Water Resources Association* 48: 74-89. Doi: 10.1111/j.1752-1688.2011.00592.x.

Shrestha RR, Osenbrück K, Rode M. 2013. Assessment of catchment response and calibration of a hydrological model using high-frequency discharge-nitrate concentration data. *Hydrology Research* 44:995-1012.Doi: 10.2166/nh.2013.087.

Sivapalan M, Takeuchi K, Franks SW, Gupta VK, Karambiri H, Lakshmi V, Liang X, McDonnell JJ, Mendiondo EM, O'Connell PE, Oki T, Pomeroy JW, Schertzer D, Uhlenbrook S, Zehe E. 2003. IAHS Decade on Predictions in Ungauged Basins (PUB), 2003–2012: Shaping an exciting future for the hydrological sciences, *Hydrological Sciences Journal* 48: 857-880. Doi: 10.1623/hysj.48.6.857.51421.

Smith VH, Tilman GD, Nekola JC. 1999. Eutrophication: impacts of excess nutrient inputs on freshwater, marine, and terrestrial ecosystems. *Environmental Pollution* 100: 179-196.

Somura H, Takeda I, Arnold JG, Mori Y, Jeong J, Kannan N, Hoffman D. 2012.Impact of suspended sediment and nutrient loading from land uses against water quality in the Hii River basin, Japan. *Journal of Hydrology* 450-451: 25-35. Doi: 10.1016/j.jhydrol.2012.05.032.

Sorooshian S, Gupta VK. 1983. Automatic calibration of conceptual rainfall-runoff models: the question of parameter observability and uniqueness. *Water Resources Research* 19: 260-268.

Stedinger JR, Vogel RM, Lee SU, Batchelder R. 2008. Appraisal of the generalized likelihood uncertainty estimation (GLUE) method. *Water Resources Research* 44, W00B06, Doi:10.1029/2008WR006822.

Strömqvist J, Arheimer B, Dahné J, Donnelly C, Lindström G. 2012. Water and nutrient predictions in ungauged basins: Set-up and evaluation of a model at the national scale. Hydrological Sciences Journal 57: 229-247.

Thiemann M, Trosset M, Gupta H. Sorooshian, S., 2001.Bayesian recursive parameter estimation for hydrologic models.*Water Resources Research* 37: 2521-2535.

Ullrich A, Volk M. 2010. Influence of different nitrate-N monitoring strategies on load estimation as a base for model calibration and evaluation. *Environmental Monitoring and Assessment* 171: 513-527. Doi: 10.1007/s10661-009-1296-8.

van der Perk, M., 1997. Effect of model structure on the accuracy and uncertainty of results from water quality models.*Hydrological Processes* 11: 227-239.

van Griensven A, Bauwens W. 2003. Multiobjective autocalibration for semidistributed water quality models. *Water Resources Research* 39: 1348. Doi: 10.1029/2003WR002284.

van Griensven A, Meixner T, Grunwald S, Bishop P, Diluzio M, Srinivasan R. 2006. A global sensitivity analysis tool for the parameters of multi-variable catchment models. *Journal of Hydrology* 324: 10-23. Doi: 10.1016/j.jhydrol.2005.09.008.

van Nieuwenhuyse EE, Jones JR. 1996. Phosphorus chlorophyll relationship in temperate streams and its variation with stream catchment area. *Canadian Journal of fisheries and aquatic sciences* 53, 99-105.

Van Rompaey AJJ, Govers G. 2002. Data quality and model complexity for regional scale soil erosion prediction. *International Journal of Geographical Information Science* 16, 663-680.

Viney, N.R., Sivapalan, M., 2001. Modelling catchment processes in the Swan-Avon river basin. *Hydrological Processes* 15(13): 2671-2685.

Viney NR, Sivapalan M, Deeley D. 2000. A conceptual model of nutrient mobilisation and transport applicable at large catchment scales. *Journal of Hydrology* 240: 23-44. Doi: 10.1016/S0022-1694(00)00320-6.

Vrugt JA, Braak CJFt, Diks CGH, Robinson BA, Hyman JM, Higdon D. 2009. Accelerating Markov Chain Monte Carlo Simulation by Differential Evolution with Self-Adaptive Randomized Subspace Sampling. *International Journal of Nonlinear Sciences & Numerical Simulation* 10: 271-288.

Vrugt JA, Gupta HV, Bastidas LA, Bouten W, Sorooshian S. 2003. Effective and efficient algorithm for multiobjective optimization of hydrologic models. *Water Resources Research* 39.

Vrugt JA, Gupta HV, Bouten W, Sorooshian S. 2003. A Shuffled Complex Evolution Metropolis algorithm for optimization and uncertainty assessment of hydrologic model parameters *Water Resources Research* 39: W00J15. Doi: 10.1029/2002WR001642.

Wade AJ, Durand P, Beaujouan V, Wessel WW, RaatKJ, Whitehead PG, Butterfield D, Rankinen K, Lepisto A. 2002. A nitrogen model for European catchments: INCA, new model structure and equations. *Hydrology and Earth System Sciences* 6: 559-582.

Wagenschein D, Rode M. 2008. Modelling the impact of river morphology on nitrogen retention—A case study of the Weisse Elster River (Germany). *Ecological Modelling* 211: 224-232.

Wang S, Zhang Z, Sun G, Strauss P, Guo J, Tang Y, Yao A. 2012. Multi-site calibration, validation, and sensitivity analysis of the MIKE SHE Model for a large watershed in northern China. *Hydrology and Earth System Sciences* 16(12): 4621-4632.

Whitehead, P., Wilson, E., Butterfield, D., 1998 a. A semi-distributed ntegrated nitrogen model for multiple source assessment in tchments (INCA): Part I — model structure and process equations. *Science of the Total Environment* 210-211: 547-558.

Whitehead P, Wilson E, Butterfield D, Seed K. 1998 b. A semi-distributed integrated flow and nitrogen model for multiple source assessment in catchments (INCA): Part II — application to large river basins in south Wales and eastern England. *Science of the Total Environment* 210-211: 559-583.

Wolock DM, Price CV. 1994. Effects of digital elevation model map scale and data resolution on a topography-based watershed model. *Water Resources Research* 30: 3041-3052.

Wondzell SM. 2011. The role of the hyporheic zone across stream networks. *Hydrological Processes* 25: 3525–3532. Doi: 10.1002/hyp.8119.

Wrede S, Seibert J, Uhlenbrook S. 2013. Distributed conceptual modelling in a Swedish lowland catchment: a multi-criteria model assessment. *Hydrology Research* 44(2): 318.

Wriedt G, Rode M. 2006. Modelling nitrate transport and turnover in a lowland catchment system. *Journal of Hydrology* 328: 157-176. DOI: 10.1016/j.jhydrol.2005.12.017.

Yapo PO, Gupta HV, Sorooshian. 1998. Multi-objective global optimization for hydrologic models. *Journal of Hydrology* 204: 83-97.

Ye S, Covino TP, Sivapalan M, Basu NB, Li HY, Wang SW. 2012. Dissolved nutrient retention dynamics in river networks: A modeling investigation of transient flows and scale effects. *Water Resources Research* 48: W00J17. Doi: 10.1029/2011WR010508.

Zawadzki IL. 1973. Errors and fluctuations of raingauge estimates of areal rainfall. *Journal of hydrology* 18: 243-255.

Zhang T. 2011. Distance-decay patterns of nutrient loading at watershed scale: Regression modeling with a special spatial aggregation strategy. *Journal of Hydrology* 402: 239-249.

Zhang X, Srinivasan R, Liew MV. 2008. Multi-site calibration of the SWAT model for hydrologic modeling. *Transactions of the ASABE* 51: 2039-2049.

Zheng Y, Wang W, Han F, Ping J. 2011. Uncertainty assessment for watershed water quality modeling: A Probabilistic Collocation Method based approach. *Advances in Water Resources* 34(7): 887-898.

Appendix 1: Hydrological nitrogen processes in the HYPE model

Appendix 1A: Model equations

(1) Snow accumulation

$$q_{SNOW} = P \cdot \partial_{SNOW}$$

$$\partial_{SNOW} = 1 - \frac{T_{CLASS} - (p_{TTMP} - p_{TINT})}{2 \cdot p_{TINT}} \quad if \quad p_{TTMP} - p_{TINT} < T_{CLASS} < p_{TTMP} + p_{TINT}$$

$$\partial_{SNOW} = 1 \quad if \quad T_{CLASS} < p_{TTMP} - p_{TINT}$$

$$\partial_{SNOW} = 0 \quad if \quad T_{CLASS} > p_{TTMP} + p_{TINT}$$

$$T_{CLASS} = T_{AIR} - p_{TCALT} \cdot c_{\Delta H}$$

(2) Snowmelt

$$q_{MELT} = \min(p_{CMLT} \cdot (T_{CLASS} - p_{TTMP}), W_{SNOW})$$

(3) Infiltration

$$q_{INF} = (P \cdot (1 - \partial_{SNOW}) + q_{MELT}) - q_{SR} - q_{MPOR}$$

(4) Surface runoff

$$q_{SR} = p_{RCSR} \cdot (P \cdot (1 - \partial_{SNOW}) + q_{MELT} - p_{THRQ}) \quad if \quad P \cdot (1 - \partial_{SNOW}) + q_{MELT} > p_{THRQ}, \quad W_{SOIL}(u) > p_{THRQ}$$

(5) Macropore flow

$$q_{MPOR} = p_{RCMP} \cdot (P \cdot (1 - \partial_{SNOW}) + q_{MELT} - p_{THRQ}) \quad if \quad P \cdot (1 - \partial_{SNOW}) + q_{MELT} > p_{THRQ}, \quad W_{SOIL}(u) > p_{THRQ\Theta}$$

(6) Saturated overland flow

$$q_{SOFL} = \max(p_{RCSOF} \cdot (W_{SOIL}(u) - \partial_{WC}(u)), 0)$$

(7) Percolation

$$q_{PERC} = \min(\max(W_{SOIL}(i) - \partial_{FC}(i), 0), \partial_{WC}(j) - W_{SOIL}(j), p_{MPERC}), \quad i = u, m \quad and \quad j = m, l$$

(8) Evapotranspiration

$q_E(i) = 0 \quad if \quad W_{SOIL}(i) - \partial_1 < 0, \quad i = u, m$

$q_E(i) = \min(\partial_{EVAP}(i) \cdot E_{POT}, W_{SOIL}(i) - \partial_1) \quad if \quad W_{SOIL}(i) - \partial_1 > p_{LP} \cdot p_{\Theta 2} \cdot c_{SLAY} \cdot 10^3, \quad i = u, m$

$q_E(i) = \min\left(\partial_{EVAP}(i) \cdot E_{POT} \cdot \dfrac{W_{SOIL}(i) - \partial_1}{p_{LP} \cdot p_{\Theta 2} \cdot c_{SLAY} \cdot 10^3}, W_{SOIL}(i) - \partial_1\right) \quad if \quad 0 < W_{SOIL}(i) - \partial_1 < p_{LP} \cdot p_{\Theta 2} \cdot c_{SLAY} \cdot 10^3,$

$i = u, m$

$E_{POT} = p_{CEVP} \cdot (T_{CLASS} - p_{TTMP}) \cdot \left(1 + p_{CEAM} \cdot \sin\left(2 \cdot \pi \cdot \dfrac{t_{DNO} - p_{CEPH}}{365}\right)\right) \quad if \quad T_{CLASS} > p_{TTMP}$

$\partial_{EVAP}(u) = \dfrac{c_{SLAY}(u) \cdot e^{(-p_{CED} \cdot c_{SLAY}(u)/2)}}{c_{SLAY}(u) \cdot e^{(-p_{CED} \cdot c_{SLAY}(u)/2)} + c_{SLAY}(m) \cdot e^{(-p_{CED} \cdot (c_{SLAY}(u) + c_{SLAY}(m)/2))}}; \quad \partial_{EVAP}(m) = 1 - \partial_{EVAP}(u)$

(9) Interflow

$q_{RUNF} = \partial_{RC} \cdot (W_{SOIL} - \partial_{FC}) \quad if \quad W_{SOIL} - \partial_{FC} > 0 \quad$ (for all soil layers above soil layer with stream)

$q_{RUNF} = \partial_{RC} \cdot ((W_{SOIL} - \partial_{FC}) - p_{\Theta 3} \cdot c_{SLAY} \cdot 10^3 \cdot (1 - c_{STRD}/c_{SLAY})) \quad if$

$((W_{SOIL} - \partial_{FC}) - p_{\Theta 3} \cdot c_{SLAY} \cdot 10^3 \cdot (1 - c_{STRD}/c_{SLAY})) > 0 \quad$ (for soil layer with stream)

$\partial_{RC}(u) = p_{RCU} + p_{RCSL} \cdot c_{SLOPE}$

$\partial_{RC}(m) = \partial_{RC}(u) \cdot e^{(-\partial_{ERC} \cdot (c_{SLAY}(u) + c_{SLAY}(m))/2)}$

$\partial_{RC}(l) = p_{RCL}$

$\partial_{ERC} = \dfrac{\ln(\partial_{RC}(u)/p_{RCL})}{c_{SLAY}(u)/2 + c_{SLAY}(m) + c_{SLAY}(l)/2}$

(10) Tile drain flow

$q_{TILE} = \max\left(\min\left(p_{RCT} \cdot \dfrac{d_{TILE}}{c_{SLAY}} \cdot \partial_3, W_{SOIL} - \partial_{FC}\right), 0\right) \quad$ (for soil layer with tile)

(11) Regional groundwater flow

$q_{GRW} = p_{RCG} \cdot (W_{SOIL}(l) - \partial_{FC})$

$Q_{GRW} = \displaystyle\sum_{\forall land\ class} q_{GRW} \cdot c_{AREA}$

(12) Soil nitrogen pool transformation processes

$$F_{DEGN} = p_{DEGN} \cdot X_{SLOWN} \cdot f(T_{SOIL}) \cdot f(\Theta)$$

$$F_{MINN} = p_{MINN} \cdot X_{FASTN} \cdot f(T_{SOIL}) \cdot f(\Theta)$$

$$F_{DENIT} = p_{DENIT} \cdot X_{IN} \cdot f(T_{SOIL}) \cdot f_2(\Theta) \cdot f(C_{IN})$$

$$f(T_{SOIL}) = \begin{cases} 2^{(T_{SOIL}-20)/10} & T_{SOIL} > 5° \\ \dfrac{2^{(T_{SOIL}-20)/10}}{5} & 0° < T_{SOIL} \le 5° \\ 0 & T_{SOIL} \le 0° \end{cases}$$

$$T_{SOIL}(t) = T_{SOIL}(t-1) \cdot \left(1 - \dfrac{1}{c_{SOILMEM} + c_{SPFROST} \cdot d_{SNOW}} - c_{WDEEP}\right) + T_{AIR} \cdot \dfrac{1}{c_{SOILMEM} + c_{SPFROST} \cdot d_{SNOW}} + c_{TDEEP} \cdot c_{WDEEP}$$

$$d_{SNOW} = \dfrac{0.1 \cdot W_{SNOW}}{p_{DENS0} + p_{DENSDT} \cdot a_{SNOW}}; \quad a_{SNOW}(t) = \dfrac{W_{SNOW}(t-1) \cdot (a_{SNOW}(t-1)+1) + q_{SNOW} \cdot 0}{W_{SNOW}(t)}$$

$$f(\Theta) = \begin{cases} p_{SATACT} & \Theta = \Theta_{SAT} \\ \min\left(\left(\dfrac{\Theta_{SAT} - \Theta}{p_{RUPP}}\right)^{p_{SMEX}}\right) \cdot (1 - p_{SATACT}) + p_{SATACT}, \left(\dfrac{\Theta - p_{\Theta 1}}{p_{RLOW}}\right)^{p_{SMEX}} & p_{\Theta 1} \le \Theta < \Theta_{SAT} \\ 0 & \Theta < p_{\Theta 1} \end{cases}$$

$$f_2(\Theta) = \begin{cases} \left(\left(\Theta/\Theta_{SAT} - p_{SMDEN}\right)/(1 - p_{SMDEN})\right)^{p_{DENEX}} & \Theta \ge p_{SMDEN} \\ 0 & \Theta < p_{SMDEN} \end{cases}$$

$$\Theta_{SAT} = p_{\Theta 1} + p_{\Theta 2} + p_{\Theta 3}$$

$$f(C_{IN}) = \dfrac{C_{IN}}{C_{IN} + p_{HSATIN}}$$

(13) Plant uptake

$$F_{PUTN} = \min(\partial_{PUT},\ 0.8 \cdot X_{IN})$$

$$\partial_{PUT} = \dfrac{p_{PUT1} \cdot p_{PUT3} \cdot \left(\dfrac{p_{PUT1} - p_{PUT2}}{p_{PUT2}}\right) \cdot e^{-p_{PUT3} \cdot (t_{DNO} - c_{BD5})}}{\left(1 + \left(\dfrac{p_{PUT1} - p_{PUT2}}{p_{PUT2}}\right) \cdot e^{-p_{PUT3} \cdot (t_{DNO} - c_{BD5})}\right)^2}$$

$$\partial_{PUT} = \partial_{PUTAUT} \quad if \quad t_{DNO} > c_{BD5}$$

$$\partial_{PUTAUT} = \dfrac{p_{PUT1} \cdot p_{PUT3} \cdot \left(\dfrac{p_{PUT1} - p_{PUT2}}{p_{PUT2}}\right) \cdot e^{-p_{PUT3} \cdot (t_{DNO} - c_{BD5} - 25)}}{\left(1 + \left(\dfrac{p_{PUT1} - p_{PUT2}}{p_{PUT2}}\right) \cdot e^{-p_{PUT3} \cdot (t_{DNO} - c_{BD5} - 25)}\right)^2} \cdot f(T_{AIR})$$

$$f(T_{AIR}) = \begin{cases} 0 & T_{AIR} < p_{TTHR} \\ \min\left(1, \left(\dfrac{1}{p_{TMAX} - p_{TTHR}}\right) \cdot T_{AIR} - \dfrac{p_{TTHR}}{p_{TMAX} - p_{TTHR}}\right) & T_{AIR} \ge p_{TTHR} \end{cases}$$

(14) Primary production, mineralization and denitification in rivers

$$F_{PRODNW} = p_{PROMIN} \cdot f(T_{WATER}) \cdot f(T_{10},T_{20}) \cdot f(\overline{C}_{TP}) \cdot V_{RIV}$$

$$F_{MINNW} = -F_{PRODNW}$$

$$F_{DENW} = p_{DENW} \cdot f(C_{IN}) \cdot f(T_{WATER}) \cdot c_{AREA}$$

$$f(T_{WATER}) = \frac{T_{WATER}}{20}; \quad T_{WATER}(t) = (1 - p_{WAIR}) \cdot T_{WATER}(t-1) + p_{WAIR} \cdot T_{AIR}$$

$$f(\overline{C}_{TP}) = \frac{\overline{C}_{TP}}{\overline{C}_{TP} + p_{HSATTP}}; \quad f(C_{IN}) = \frac{C_{IN}}{C_{IN} + p_{HSATIN}} \quad f(T_{10},T_{20}) = \frac{T_{10} - T_{20}}{5}$$

Appendix 1B: Notation

State variables

C_{IN} concentration of inorganic nitrogen (mg/l)
V_{RIV} volume of river (m³)
W_{SNOW} soil moisture (mm)
W_{SOIL} water content of soil (mm)
X_{FASTN} fast turnover N pool (kg/km²)
X_{IN} inorganic N pool (kg/km²)
X_{SLOWN} slow turnover N pool (kg/km²)

Process variables

F_{DEGN} degradation of humusN (kg/km² d)
F_{DENIT} denitrification of soil (kg/km² d)
F_{DENW} denitrification in rivers (kg/d)
F_{MINN} mineralization of fastN (kg/km² d)
F_{MINNW} net mineralization of N in water (kg/d)
F_{PRODNW} net primary production of N in water (kg/d)
F_{PUTN} plant uptake of N (kg/km² d)
q_E evapotranspiration (mm)
q_{GRW} regional groundwater outflow from a sub-basin (mm/d)
Q_{GRW} regional groundwater outflow from a sub-basin (mm m²/d)
q_{INF} infiltration (mm/d)
q_{MELT} snowmelt (mm/d)
q_{MPOR} macropore flow (mm/d)
q_{PERC} percolation (mm/d)
q_{RUNF} soil runoff (mm/d)
q_{SNOW} snow fall (mm/d)
q_{SOFL} saturated overland flow (mm/d)
q_{SR} surface runoff due to excess infiltration (mm/d)
q_{TILE} tile drainage runoff (mm/d)

Other variables

a_{SNOW} age of snow (d)

$\overline{C}_{TP}$ mean total phosphorus contention (mg/l)

d_{SNOW} snow depth (cm)

d_{TILE} water storage above the tile depth (m)

E_{POT} potential evapotranspiration (mm/d)

i soil layer index (u, m or l)

j soil layer index (u, m or l)

l lowest soil layer (-)

m second soil layer (-)

P precipitation(mm/d)

t time step (d)

T_{10} average water temperature over a 10 day period (°C)

T_{20} average water temperature over a 20 day period (°C)

T_{AIR} air temperature for sub-basin (°C)

T_{CLASS} air temperature for class (°C)

t_{DNO} day number of year (d)

T_{SOIL} temperature of soil (°C)

T_{WATER} water temperature (°C)

W_{SOIL} water content of soil (mm)

u lowest soil layer (-)

∂_1 maximum water content not available for evapotranspiration (mm)

∂_3 maximum water content available for runoff (mm)

∂_{ERC} exponential rate of runoff coefficient (1/m)

∂_{EVAP} fraction of evapotranspiration from soil layer (-)

∂_{FC} water content threshold for runoff (mm)

∂_{PUT} potential plant uptake (g/m²d)

∂_{PUTAUT} potential plant uptake in autumn for autumn crop (kg/km² d)

∂_{RC} soil runoff coefficient (1/d)

∂_{SNOW} snow fraction of precipitation (-)

∂_{WC} maximum water content of soil (mm)

Θ water content of soil (-)

Θ_{SAT} water content at saturation (-)

Model parameters

p_{CEAM} amplitude of evapotranspiration seasonal correction (general) (-)

p_{CED} decrease of evapotranspiration with soil depth (general) (1/m)

p_{CEPH} phase of evapotranspiration seasonal correction (general) (d)

p_{CEVP} rate of potential evapotranspiration (land use) (mm/d °C)

p_{CMLT} snow melt coefficient (land use) (mm/d°C)

p_{DEGN} degradation of humus N to fast N (general) (1/d)

p_{DENEX} exponent in soil moisture function for denitrification (general) (-)

p_{DENIT} denitrification rate in soil (general) (1/d)

p_{DENS0} snow density of new snow (general) (-)

p_{DENSDT} change of snow density with time (general) (1/d)

p_{DENW} denitrification in water (kg/m^2 d)

p_{HSATIN} half saturation point for IN concentration (general) (mg/l)

p_{HSATTP} half saturation point for TP concentration (general) (mg/l)

p_{LP} limit for potential evapotranspiration (general) (-)

p_{MINN} mineralization of fast N to IN (general) (1/d)

p_{MPERC} maximum percolation (soil type) (mm/d)

$p_{PRODMIN}$ primary production mineralization parameter (general) (kg/m^3 d)

p_{PUT1} plant uptake parameter (crop type) (g/m^2)

p_{PUT2} plant uptake parameter (crop type) (g/m^2)

p_{PUT3} plant uptake parameter (crop type) (1/d)

p_{RCG} runoff coefficient for regional groundwater flow (general) (1/d)

p_{RCL} soil runoff coefficient for lowest layer (soil type) (1/d)

p_{RCMP} runoff coefficient for macropore flow (soil type) (-)

p_{RCSL} runoff coefficient dependence on slope (soil type) (1/d %)

p_{RCSOF} runoff coefficient for saturation overland flow (land use) (1/d)

p_{RCSR} runoff coefficient for surface runoff (soil type) (-)

p_{RCT} tile drainage runoff coefficient (soil type) (1/d)

p_{RCU} soil runoff coefficient for top layer (soil type) (1/d)

p_{RLOW} lower range in soil moisture function (general) (-)

p_{RUPP} upper range in soil moisture function (general) (-)

p_{SATACT} activity at saturation (general) (-)

p_{SMDEN} coefficient in soil moisture function for denitrification (general) (-)

p_{SMEX} exponent in soil moisture function (general) (-)

p_{TCALT} air temperature's elevation dependence (general) (°C/m)

p_{THRQ} soil moisture threshold for surface runoff and macropre flow (soil type) (mm/d)

$p_{THR\Theta}$ soil moisture threshold for surface runoff and macropre flow (soil type) (mm)

p_{TINT} half temperature interval for mixed snow and rain (general) (°C)

p_{TMAX} the parameter in winter crop nutrient uptake calculation (°C)

p_{TTMP} temperature threshold (land use) (°C)

p_{TTHR} temperature threshold for plant nutrient uptake during late autumn (general) (°C)

p_{WAIR} weight for air temperature in water temperature Equation (general) (-)

$p_{\Theta 1}$ fraction of soil layer where water is not available for evapotranspiration (soil type) (-)

$p_{\Theta 2}$ fraction of soil layer where water is available for evapotranspiration but not for runoff (soil type) (-)

$p_{\Theta 3}$ fraction of soil layer where water is available for runoff (soil type) (-)

Constants/input data

c_{AREA} class area (m^2)

c_{BD2} day number of sowing date in spring (d)

c_{BD5} day number of sowing date in autumn (d)

c_{SLAY} soil layer thickness (m)

c_{SLOPE} slope of sub-basin (%)

$c_{SOILMEM}$ soil temperature memory (30d)

$c_{SPFROST}$ soil temperature snow dependence (10 d/cm)

c_{STRD} stream depth below soil surface (m)

c_{TDEEP} deep soil temperature (5°C)

c_{TILE} depth of tile and drainage pipe (m)

c_{WDEEP} deep soil temperature weight (0.001)

$c_{\Delta H}$ class's elevation deviation from sub-basin mean elevation (m)

Appendix 2: How PEST does optimization, sensitivity analysis and predictive analysis

Appendix 2A: parameter optimization

At the beginning of each iteration the relationship between model parameters and model-generated observations is linearized by formulating it as a Taylor expansion about the currently best parameter set; hence the derivatives of all observations with respect to all parameters must be calculated. This linearized problem is then solved for a better parameter set, and the new parameters tested by running the model again. By comparing parameter changes and objective function improvement achieved through the current iteration with those achieved in previous iterations, PEST can tell whether it is worth undertaking another optimization iteration; if so the whole process is repeated (Doherty, 2005). The optimum is reached, when the gradient becomes small with respect to a certain tolerance limit. Derivatives of observations with respect to parameters are calculated using finite differences. The ability to calculate the derivatives of all observations with respect to all adjustable parameters is fundamental to the Gauss-Marquardt-Levenberg method of parameter estimation; these derivatives are stored as the elements of the Jacobian matrix used for sensitivity analysis. The objective function used is the weighted sum of squared ϕ (Doherty, 2005; Rode et al., 2007 b)

$$\phi = \sum_{i=1}^{m} (\omega_i r_i)^2$$

where ω_i is the weight attached to the i th observation, m is the number of observations and r_i is the difference between the model outcome and the corresponding observation (i.e. the i th residual of discharge/nitrogen concentration simulations in our study). The weights assigned are inversely proportional to the standard deviations of the observed values. In this way, the observations with different magnitude can have similar contribution to the objective function, which is important, especially in multi-objective calibration (e.g. calibrating hydrological and water parameters simultaneously in integrated hydrological water quality modeling).

Appendix 2B: Sensitivity analysis

The composite sensitivity of each parameter is the normalized (with respect to the number of observations) magnitude of the column of the Jacobian matrix pertaining to that parameter. The Jacobian matrix comprised of m rows (one for each observation), and the n elements of each row are the derivatives of one particular observation with respect to each of the n parameters. In fact, most of the time consumed during each PEST optimization iteration is devoted to calculation of the Jacobian matrix. During this process the model must be run at least n times, where n is the number of adjustable parameters. Immediately after it calculates the Jacobian matrix, PEST writes composite parameter sensitivities to the output file.

Composite parameter sensitivities are useful for identifying those parameters, which may degrade the performance of the parameter estimation process due to low sensitivity to model outcomes. The use of relative sensitivities in addition to normal sensitivities assists in comparing the effects that different parameters have on the parameter estimation process when these parameters are of different type, and possibly of very different magnitudes (Bahremand and De Smedt, 2010; Doherty, 2005; Rode et al., 2007 b).

Appendix 2C: Predictive analysis

To run PEST in predictive analysis mode the user informs PEST of the objective function value (Φ_0) below which the model can be considered to be calibrated; this value is normally just slightly above the minimum objective function value (Φ_{min}) as determined in a previous PEST calibration run and Φ_0 is expresses as Φ_{min}+ δ (δ is relatively small). To do predictive analysis by changing parameters (the way PEST does it), they must be varied in such a way that the objective function hardly changes (i.e. keeping the model under calibration conditions) and the model can still be assumed to be behavioral. To construct a 95% confidence intervals of the predictions, the δ for calibrated state is calculated following the formula given below (Equation 2.2) where α equals 0.05 (Doherty and Johnston, 2003; Lin and Radcliffe, 2006):

$$\delta = n\sigma^2 F_\alpha(n, m-n)$$

where σ^2 is given by equation $\sigma^2 = \phi_{min}/(m-n)$, in which n is the number of parameters requiring estimation, m is the number of the observations comprising the calibration dataset, and F(.,.) denotes the F-distribution.

When a model is set up for predictive analysis, a composite model is constructed that is comprised of the model run under calibration conditions followed by the model run under predictive conditions. There can be as many field observations corresponding to the former model component as desired, these being the "calibration observations". However, there should be only one output for the predictive model component, which is used by PEST as a target observation. This observation, to be recognized by PEST, should be the sole member of an observation group named "predict". It is important to note that PEST takes no notice of either the "observed value" of this observation or of the weight assigned to this observation.

PEST's job is simply to raise or lower the model output corresponding to this observation, while maintaining the objective function at or below $\Phi_{min}+\delta$ (Bahremand and De Smedt, 2010; Doherty, 2005).

Monte-Carlo analysis is often used to examine uncertainty in model predictions. Parameter sets can be generated at random; for each such parameter set the model is run under calibration conditions. If the resulting objective function is above $\Phi_{min}+\delta$ the parameter set is rejected. If it is below $\Phi_{min}+\delta$ the model is then run under predictive conditions. After thousands of model runs have been undertaken a suite of predictions will have been built up, all generated by parameter sets which satisfy calibration constraints. In many cases a probability distribution can then be attached to these predictions, based on where corresponding calibration objective functions lie with respect to Φ_{min} and $\Phi_{min}+\delta$ contour. This method of predictive analysis has many advantages; however, its main disadvantage is in the number of model runs required. Where there are any more than a few adjustable parameters, the dimensionality of the problem requires that millions of model runs be undertaken, rendering the method intractable in many practical settings. Markov Chain Monte Carlo (MCMC) methods have the ability to explore post-calibration parameter and predictive uncertainty with much higher efficiency than the basic Monte Carlo method described above. However the cost in model runs is still extremely high, especially where parameters' number more than just a few, and correlation between parameters is high (Doherty, 2005).

Appendix 3: Bayesian inference of the posterior probability density function of hydrologic model parameters using DREAM$_{(ZS)}$

DREAM$_{(ZS)}$ is based on original DREAM algorithm, but uses sampling from archive of past states to generate candidate points in each individual chain. DREAM uses DE-MC (Differential Evolution-Markov Chain) as its main building block. The subscript for generation time t is dropped. In the following, the current state of the ith chain is given by a d-dimensional vector X^i $(i = 1,...,N)$ and its jth element by x_j^i. The DREAM algorithm is explained as follows (after Vrugt et al., 2009):

1. Draw an initial population $\{X^i, i = 1,...,N\}$ using the prior distribution.

2. Compute the density $\pi(X^i)$ for $i = 1,...,N$.

   FOR $i \leftarrow 1,...,N$ DO (CHAIN EVOLUTION)

3. Generate a candidate point, Z^i in chain i.

$$Z^i = X^i + (1_d + e)\gamma(\delta, d')\left[\sum_{j=1}^{\delta} X^{r_1(j)} - \sum_{n=1}^{\delta} X^{r_2(n)}\right] + \varepsilon$$

where δ signifies the number of pairs used to generate the proposal, and $r_1(j)$, $r_2(n) \in \{1,...,N\}$; $r_1(j) \neq r_2(n) \neq i$ for $j = 1,...,\delta$ and $n = 1,...,\delta$. The values of e and ε are drawn from $U_d(-b, b)$ and $N_d(0, b^*)$ with $|b| < 1$, and b^* small compared to the width of the target distribution, respectively, and the value of jump-size γ depends on δ and d', the number of dimensions that will be updated jointly.

4. Replace each element $(j = 1,...,d)$ of proposal z_j^i with x_j^i using a binomial scheme with probability $1 - CR$, where CR is the crossover probability. With $CR = 1$, all dimensions are updated jointly and $d' = d$.

5. Compute $\pi(Z^i)$ and $\partial(X^i, Z^i)$ of the candidate point.

6. If accepted, $X^i = Z^i$, otherwise remain at X^i.

   END FOR (CHAIN EVOLUTION)

7. Remove outlier chains using the Inter QuartileRange (IQR) statistic. This is done during burnin.

8. Compute the Gelman-Rubin $\hat{R}_j$ convergence diagnostic for each dimension $j = 1,...,d$ using the last 50% of the samples in each chain.

9. If $\hat{R}_j < 1.2$ for all j, stop, otherwise go to CHAIN EVOLUTION.

i want morebooks!

Buy your books fast and straightforward online - at one of the world's fastest growing online book stores! Environmentally sound due to Print-on-Demand technologies.

Buy your books online at
www.get-morebooks.com

Kaufen Sie Ihre Bücher schnell und unkompliziert online – auf einer der am schnellsten wachsenden Buchhandelsplattformen weltweit!
Dank Print-On-Demand umwelt- und ressourcenschonend produziert.

Bücher schneller online kaufen
www.morebooks.de

OmniScriptum Marketing DEU GmbH
Heinrich-Böcking-Str. 6-8
D - 66121 Saarbrücken
Telefax: +49 681 93 81 567-9

info@omniscriptum.de
www.omniscriptum.de

Printed by Books on Demand GmbH, Norderstedt / Germany